Practical Machinery Vibration Analysis and Predictive Maintenance

T0229124

Practical Machinery Vibration Analysis and Predictive Maintenance

Paresh Girdhar BEng (Mech. Eng), Girdhar and Associates

Edited by
C. Scheffer PhD, MEng, SAIMechE

Series editor: Steve Mackay

AMSTERDAM • BOSTON • HEIDELBERG • LONDON
NEW YORK • OXFORD • PARIS • SAN DIEGO
SAN FRANCISCO • SINGAPORE • SYDNEY • TOKYO
Newnes is an imprint of Elsevier

Newnes
An imprint of Elsevier
Linacre House, Jordan Hill, Oxford OX2 8DP
200 Wheeler Road, Burlington, MA 01803

First published 2004

British Library Cataloguing in Publication Data
Girdhar, P.
    Practical machinery vibration analysis and predictive
    maintenance. – (Practical professional)
    1. Machinery – Vibration   2. Vibration – Measurement
    3. Machinery – Maintenance and repair
    I. Title
    621.8'11

Library of Congress Cataloguing in Publication Data
A catalogue record for this book is available from the Library of Congress

ISBN   978-0-7506-6275-8

For information on all Newnes Publications
visit our website at www.newnespress.com

Typeset and edited by Integra Software Services Pvt. Ltd, Pondicherry, India
www.integra-india.com

Transferred to Digital Printing in 2008

Contents

Preface

This practical book provides a detailed examination of the detection, location and diagnosis of faults in rotating and reciprocating machinery using vibration analysis. The basics and underlying physics of vibration signals are first examined. The acquisition and processing of signals are then reviewed followed by a discussion of machinery fault diagnosis using vibration analysis. Hereafter the important issue of rectifying faults that have been identified using vibration analysis is covered. The book is concluded by a review of the other techniques of predictive maintenance such as oil and particle analysis, ultrasound and infrared thermography. The latest approaches and equipment used together with current research techniques in vibration analysis are also highlighted in the text.

We would hope that you will gain the following from this book:

- An understanding of the basics of vibration measurement
- The basics of signal analysis
- Understanding the measurement procedures and the characteristics of vibration signals
- Ability to use Data Acquisition equipment for vibration signals
- How to apply vibration analysis for different machinery faults
- How to apply specific techniques for pumps, compressors, engines, turbines and motors
- How to apply vibration based fault detection and diagnostic techniques
- The ability to diagnose simple machinery related problems with vibration analysis techniques
- How to apply advanced signal processing techniques and tools to vibration analysis
- How to detect, locate and diagnose faults in rotating and reciprocating machinery using vibration analysis techniques
- Ability to identify conditions of resonance and be able to rectify these problems
- How to apply basic allied predictive techniques such as oil analysis, thermography, ultrasonics and performance evaluation.

Typical people who will find this book useful include:

- Instrumentation & Control Engineers
- Maintenance Engineers
- Mechanical Engineers & Technicians
- Control Technicians
- Electrical Engineers
- Electricians
- Maintenance Engineers & Technicians
- Process Engineers
- Consulting Engineers
- Automation Engineers.

1

Predictive maintenance techniques: Part 1

Predictive maintenance basics

1.1 Maintenance philosophies

If we were to do a survey of the maintenance philosophies employed by different process plants, we would notice quite a bit of similarity despite the vast variations in the nature of their operations. These maintenance philosophies can usually be divided into four different categories:

- Breakdown or run to failure maintenance
- Preventive or time-based maintenance
- Predictive or condition-based maintenance
- Proactive or prevention maintenance.

These categories are briefly described in Figure 1.1.

1.1.1 Breakdown or run to failure maintenance

The basic philosophy behind breakdown maintenance is to allow the machinery to run to failure and only repair or replace damaged components just before or when the equipment comes to a complete stop. This approach works well if equipment shutdowns do not affect production and if labor and material costs do not matter.

The disadvantage is that the maintenance department perpetually operates in an unplanned 'crisis management' mode. When unexpected production interruptions occur, the maintenance activities require a large inventory of spare parts to react immediately. Without a doubt, it is the most inefficient way to maintain a production facility. Futile attempts are made to reduce costs by purchasing cheaper spare parts and hiring casual labor that further aggravates the problem.

The personnel generally have a low morale in such cases as they tend to be overworked, arriving at work each day to be confronted with a long list of unfinished work and a set of new emergency jobs that occurred overnight.

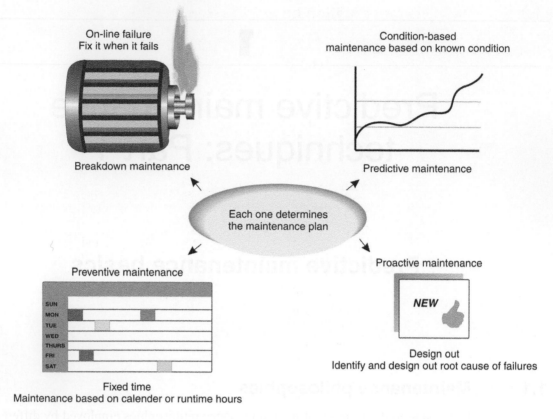

Figure 1.1
Maintenance Philosophies

Despite the many technical advances in the modern era, it is still not uncommon to find production plants that operate with this maintenance philosophy.

1.1.2   Preventive or time-based maintenance

The philosophy behind preventive maintenance is to schedule maintenance activities at predetermined time intervals, based on calendar days or runtime hours of machines. Here the repair or replacement of damaged equipment is carried out before obvious problems occur. This is a good approach for equipment that does not run continuously, and where the personnel have enough skill, knowledge and time to perform the preventive maintenance work.

The main disadvantage is that scheduled maintenance can result in performing maintenance tasks too early or too late. Equipment would be taken out for overhaul at a certain number of running hours. It is possible that, without any evidence of functional failure, components are replaced when there is still some residual life left in them. It is therefore quite possible that reduced production could occur due to unnecessary maintenance. In many cases, there is also a possibility of diminished performance due to incorrect repair methods. In some cases, perfectly good machines are disassembled, their good parts removed and discarded, and new parts are improperly installed with troublesome results.

1.1.3   Predictive or condition-based maintenance

This philosophy consists of scheduling maintenance activities only when a functional failure is detected.

Mechanical and operational conditions are periodically monitored, and when unhealthy trends are detected, the troublesome parts in the machine are identified and scheduled for maintenance. The machine would then be shut down at a time when it is most convenient, and the damaged components would be replaced. If left unattended, these failures could result in costly secondary failures.

One of the advantages of this approach is that the maintenance events can be scheduled in an orderly fashion. It allows for some lead-time to purchase parts for the necessary repair work and thus reducing the need for a large inventory of spares. Since maintenance work is only performed when needed, there is also a possible increase in production capacity.

A possible disadvantage is that maintenance work may actually increase due to an incorrect assessment of the deterioration of machines. To track the unhealthy trends in vibration, temperature or lubrication requires the facility to acquire specialized equipment to monitor these parameters and provide training to personnel (or hire skilled personnel). The alternative is to outsource this task to a knowledgeable contractor to perform the machine-monitoring duties.

If an organisation had been running with a breakdown or preventive maintenance philosophy, the production team and maintenance management must both conform to this new philosophy.

It is very important that the management supports the maintenance department by providing the necessary equipment along with adequate training for the personnel. The personnel should be given enough time to collect the necessary data and be permitted to shut down the machinery when problems are identified.

1.1.4   Proactive or prevention maintenance

This philosophy lays primary emphasis on tracing all failures to their root cause. Each failure is analyzed and proactive measures are taken to ensure that they are not repeated. It utilizes all of the predictive/preventive maintenance techniques discussed above in conjunction with root cause failure analysis (RCFA). RCFA detects and pinpoints the problems that cause defects. It ensures that appropriate installation and repair techniques are adopted and implemented. It may also highlight the need for redesign or modification of equipment to avoid recurrence of such problems.

As in the predictive-based program, it is possible to schedule maintenance repairs on equipment in an orderly fashion, but additional efforts are required to provide improvements to reduce or eliminate potential problems from occurring repeatedly.

Again, the orderly scheduling of maintenance allows lead-time to purchase parts for the necessary repairs. This reduces the need for a large spare parts inventory, because maintenance work is only performed when it is required. Additional efforts are made to thoroughly investigate the cause of the failure and to determine ways to improve the reliability of the machine. All of these aspects lead to a substantial increase in production capacity.

The disadvantage is that extremely knowledgeable employees in preventive, predictive and prevention/proactive maintenance practices are required. It is also possible that the work may require outsourcing to knowledgeable contractors who will have to work closely with the maintenance personnel in the RCFA phase. Proactive maintenance also requires procurement of specialized equipment and properly trained personnel to perform all these duties.

1.2   Evolution of maintenance philosophies

Machinery maintenance in industry has evolved from breakdown maintenance to time-based preventive maintenance. Presently, the predictive and proactive maintenance philosophies are the most popular.

Breakdown maintenance was practiced in the early days of production technology and was reactive in nature. Equipment was allowed to run until a functional failure occurred. Secondary damage was often observed along with a primary failure.

This led to time-based maintenance, also called preventive maintenance. In this case, equipment was taken out of production for overhaul after completing a certain number of running hours, even if there was no evidence of a functional failure. The drawback of this system was that machinery components were being replaced even when there was still some functional lifetime left in them. This approach unfortunately could not assist to reduce maintenance costs.

Due to the high maintenance costs when using preventive maintenance, an approach to rather schedule the maintenance or overhaul of equipment based on the condition of the equipment was needed. This led to the evolution of predictive maintenance and its underlying techniques.

Predictive maintenance requires continuous monitoring of equipment to detect and diagnose defects. Only when a defect is detected, the maintenance work is planned and executed.

Today, predictive maintenance has reached a sophisticated level in industry. Till the early 1980s, justification spreadsheets were used in order to obtain approvals for condition-based maintenance programs. Luckily, this is no longer the case.

The advantages of predictive maintenance are accepted in industry today, because the tangible benefits in terms of early warnings about mechanical and structural problems in machinery are clear. The method is now seen as an essential detection and diagnosis tool that has a certain impact in reducing maintenance costs, operational vs repair downtime and inventory hold-up.

In the continuous process industry, such as oil and gas, power generation, steel, paper, cement, petrochemicals, textiles, aluminum and others, the penalties of even a small amount of downtime are immense. It is in these cases that the adoption of the predictive maintenance is required above all.

Through the years, predictive maintenance has helped improve productivity, product quality, profitability and overall effectiveness of manufacturing plants.

Predictive maintenance in the actual sense is a philosophy – an attitude that uses the actual operating conditions of the plant equipment and systems to optimize the total plant operation.

It is generally observed that manufacturers embarking upon a predictive maintenance program become more aware of the specific equipment problems and subsequently try to identify the root causes of failures. This tendency led to an evolved kind of maintenance called proactive maintenance.

In this case, the maintenance departments take additional time to carry out precision balancing, more accurate alignments, detune resonating pipes, adhere strictly to oil check/change schedules, etc. This ensures that they eliminate the causes that may give rise to defects in their equipment in the future.

This evolution in maintenance philosophy has brought about longer equipment life, higher safety levels, better product quality, lower life cycle costs and reduced emergencies and panic decisions precipitated by major and unforeseen mechanical failures.

Putting all this objectively, one can enumerate the benefits in the following way:

- *Increase in machine productivity:* By implementing predictive maintenance, it may be possible to virtually eliminate plant downtime due to unexpected equipment failures.
- *Extend intervals between overhauls:* This maintenance philosophy provides information that allows scheduling maintenance activities on an 'as needed' basis.
- *Minimize the number of 'open, inspect and repair if necessary' overhaul routines:* Predictive maintenance pinpoints specific defects and can thus make maintenance work more focused, rather than investigating all possibilities to detect problems.
- *Improve repair time:* Since the specific equipment problems are known in advance, maintenance work can be scheduled. This makes the maintenance work faster and smoother. As machines are stopped before breakdowns occur, there is virtually no secondary damage, thus reducing repair time.
- *Increase machine life:* A well-maintained machine generally lasts longer.
- *Resources for repair can be properly planned:* Prediction of equipment defects reduces failure detection time, thus also failure reporting time, assigning of personnel, obtaining the correct documentation, securing the necessary spares, tooling and other items required for a repair.
- *Improve product quality:* Often, the overall effect of improved maintenance is improved product quality. For instance, vibration in paper machines has a direct effect on the quality of the paper.
- *Save maintenance costs:* Studies have shown that the implementation of a proper maintenance plan results in average savings of 20–25% in direct maintenance costs in conjunction with twice this value in increased production.

1.3 Plant machinery classification and recommendations

1.3.1 Maintenance strategy

The above-mentioned maintenance philosophies have their own advantages and disadvantages and are implemented after carrying out a criticality analysis on the plant equipment. Usually the criticality analysis categorizes the equipment as:

- Critical
- Essential
- General purpose.

The *critical* equipment are broadly selected on the following basis:

- If their failure can affect plant safety.
- Machines that are essential for plant operation and where a shutdown will curtail the production process.
- Critical machines include unspared machinery trains and large horsepower trains.
- These machines have high capital cost, they are very expensive to repair (e.g., high-speed turbomachinery) or take a long time to repair.

- Perennial 'bad actors' or machines that wreck on the slightest provocation of an off-duty operation.
- Finally, machinery trains where better operation could save energy or improve production.

In all probability, the proactive and predictive maintenance philosophy is adopted for critical equipment. Vibration-monitoring instruments are provided with continuous, full-time monitoring capabilities for these machines. Some systems are capable of monitoring channels simultaneously so that rapid assessment of the entire machine train is possible.

The *essential* equipment are broadly selected on the following basis:

- Failure can affect plant safety.
- Machines that are essential for plant operation and where a shutdown will curtail a unit operation or a part of the process.
- They may or may not have an installed spare available.
- Start-up is possible but may affect production process.
- High horsepower or high speed but might not be running continuously.
- Some machines that demand time-based maintenance, like reciprocating compressors.
- These machines require moderate expenditure, expertise and time to repair.
- Perennial 'bad actors' or machines that wreck at a historically arrived time schedule. For example, centrifugal fans in corrosive service.

In many cases, the preventive maintenance philosophy, and at times even a less sophisticated predictive maintenance program is adopted for such equipment. These essential machines do not need to have the same monitoring instrumentation requirements as critical machines. Vibration-monitoring systems installed on essential machines can be of the scanning type, where the system switches from one sensor to the next to display the sensor output levels one by one.

The *general purpose* equipment are broadly selected on the following basis:

- Failure does not affect plant safety.
- Not critical to plant production.
- Machine has an installed spare or can operate on demand.
- These machines require low to moderate expenditure, expertise and time to repair.
- Secondary damage does not occur or is minimal.

Usually it is acceptable to adopt the breakdown maintenance philosophy on general purpose equipment. However, in modern plants, even general purpose machines are not left to chance.

These machines do not qualify them for permanently installed instrumentation or a continuous monitoring system. They are usually monitored with portable instruments.

1.4   Principles of predictive maintenance

Predictive maintenance is basically a condition-driven preventive maintenance. Industrial or in-plant average life statistics are not used to schedule maintenance activities in this case. Predictive maintenance monitors mechanical condition, equipment efficiency and other parameters and attempts to derive the approximate time of a functional failure.

A comprehensive predictive maintenance program utilizes a combination of the most cost-effective tools to obtain the actual operating conditions of the equipment and plant systems. On the basis of this collected data, the maintenance schedules are selected.

Predictive maintenance uses various techniques such as vibration analysis, oil and wear debris analysis, ultrasonics, thermography, performance evaluation and other techniques to assess the equipment condition.

Predictive maintenance techniques actually have a very close analogy to medical diagnostic techniques. Whenever a human body has a problem, it exhibits a symptom. The nervous system provides the information – this is the detection stage. Furthermore, if required, pathological tests are done to diagnose the problem. On this basis, suitable treatment is recommended (see Figure 1.2).

Figure 1.2
Predictive maintenance

In a similar way, defects that occur in a machine always exhibit a symptom in the form of vibration or some other parameter. However, this may or may not be easily detected on machinery systems with human perceptions.

It is here that predictive maintenance techniques come to assistance. These techniques detect symptoms of the defects that have occurred in machines and assist in diagnosing the exact defects that have occurred. In many cases, it is also possible to estimate the severity of the defects.

The specific techniques used depend on the type of plant equipment, their impact on production or other key parameters of plant operation. Of further importance are the goals and objectives that the predictive maintenance program needs to achieve.

1.5   Predictive maintenance techniques

There are numerous predictive maintenance techniques, including:

    (a) *Vibration monitoring:*   This is undoubtedly the most effective technique to detect mechanical defects in rotating machinery.

    (b) *Acoustic emission:*   This can be used to detect, locate and continuously monitor cracks in structures and pipelines.

    (c) *Oil analysis:*   Here, lubrication oil is analyzed and the occurrence of certain microscopic particles in it can be connected to the condition of bearings and gears.

(d) *Particle analysis:* Worn machinery components, whether in reciprocating machinery, gearboxes or hydraulic systems, release debris. Collection and analysis of this debris provides vital information on the deterioration of these components.

(e) *Corrosion monitoring:* Ultrasonic thickness measurements are conducted on pipelines, offshore structures and process equipment to keep track of the occurrence of corrosive wear.

(f) *Thermography:* Thermography is used to analyze active electrical and mechanical equipment. The method can detect thermal or mechanical defects in generators, overhead lines, boilers, misaligned couplings and many other defects. It can also detect cell damage in carbon fiber structures on aircrafts.

(g) *Performance monitoring:* This is a very effective technique to determine the operational problems in equipment. The efficiency of machines provides a good insight on their internal conditions.

Despite all these methods, it needs to be cautioned that there have been cases where predictive maintenance programs were not able to demonstrate tangible benefits for an organisation. The predominant causes that lead to failure of predictive maintenance are inadequate management support, bad planning and lack of skilled and trained manpower.

Upon activating a predictive maintenance program, it is very essential to decide on the specific techniques to be adopted for monitoring the plant equipment. The various methods are also dependent on type of industry, type of machinery and also to a great extent on availability of trained manpower.

It is also necessary to take note of the fact that predictive maintenance techniques require technically sophisticated instruments to carry out the detection and diagnostics of plant machinery. These instruments are generally very expensive and need technically competent people to analyze their output.

The cost implications, whether on sophisticated instrumentation or skilled manpower, often lead to a question mark about the plan of adopting predictive maintenance philosophy.

However, with management support, adequate investments in people and equipment, predictive maintenance can yield very good results after a short period of time.

1.6   Vibration analysis – a key predictive maintenance technique

1.6.1   Vibration analysis (detection mode)

Vibration analysis is used to determine the operating and mechanical condition of equipment. A major advantage is that vibration analysis can identify developing problems before they become too serious and cause unscheduled downtime. This can be achieved by conducting regular monitoring of machine vibrations either on continuous basis or at scheduled intervals.

Regular vibration monitoring can detect deteriorating or defective bearings, mechanical looseness and worn or broken gears. Vibration analysis can also detect misalignment and unbalance before these conditions result in bearing or shaft deterioration.

Trending vibration levels can identify poor maintenance practices, such as improper bearing installation and replacement, inaccurate shaft alignment or imprecise rotor balancing.

All rotating machines produce vibrations that are a function of the machine dynamics, such as the alignment and balance of the rotating parts. Measuring the amplitude of vibration at certain frequencies can provide valuable information about the accuracy of shaft alignment and balance, the condition of bearings or gears, and the effect on the machine due to resonance from the housings, piping and other structures.

Vibration measurement is an effective, non-intrusive method to monitor machine condition during start-ups, shutdowns and normal operation. Vibration analysis is used primarily on rotating equipment such as steam and gas turbines, pumps, motors, compressors, paper machines, rolling mills, machine tools and gearboxes.

Recent advances in technology allow a limited analysis of reciprocating equipment such as large diesel engines and reciprocating compressors. These machines also need other techniques to fully monitor their operation.

A vibration analysis system usually consists of four basic parts:

1. Signal pickup(s), also called a transducer
2. A signal analyzer
3. Analysis software
4. A computer for data analysis and storage.

These basic parts can be configured to form a continuous online system, a periodic analysis system using portable equipment, or a multiplexed system that samples a series of transducers at predetermined time intervals.

Hard-wired and multiplexed systems are more expensive per measurement position. The determination of which configuration would be more practical and suitable depends on the critical nature of the equipment, and also on the importance of continuous or semi-continuous measurement data for that particular application.

1.6.2   Vibration analysis (diagnosis mode)

Operators and technicians often detect unusual noises or vibrations on the shop floor or plant where they work on a daily basis. In order to determine if a serious problem actually exists, they could proceed with a vibration analysis. If a problem is indeed detected, additional spectral analyses can be done to accurately define the problem and to estimate how long the machine can continue to run before a serious failure occurs.

Vibration measurements in analysis (diagnosis) mode can be cost-effective for less critical equipment, particularly if budgets or manpower are limited. Its effectiveness relies heavily on someone detecting unusual noises or vibration levels. This approach may not be reliable for large or complex machines, or in noisy parts of a plant. Furthermore, by the time a problem is noticed, a considerable amount of deterioration or damage may have occurred.

Another application for vibration analysis is as an acceptance test to verify that a machine repair was done properly. The analysis can verify whether proper maintenance was carried out on bearing or gear installation, or whether alignment or balancing was done to the required tolerances. Additional information can be obtained by monitoring machinery on a periodic basis, for example, once per month or once per quarter. Periodic analysis and trending of vibration levels can provide a more subtle indication of bearing or gear deterioration, allowing personnel to project the machine condition into the foreseeable future. The implication is that equipment repairs can be planned to commence during normal machine shutdowns, rather than after a machine failure has caused unscheduled downtime.

1.6.3   Vibration analysis – benefits

Vibration analysis can identify improper maintenance or repair practices. These can include improper bearing installation and replacement, inaccurate shaft alignment or imprecise rotor balancing. As almost 80% of common rotating equipment problems are related to misalignment and unbalance, vibration analysis is an important tool that can be used to reduce or eliminate recurring machine problems.

Trending vibration levels can also identify improper production practices, such as using equipment beyond their design specifications (higher temperatures, speeds or loads). These trends can also be used to compare similar machines from different manufacturers in order to determine if design benefits or flaws are reflected in increased or decreased performance.

Ultimately, vibration analysis can be used as part of an overall program to significantly improve equipment reliability. This can include more precise alignment and balancing, better quality installations and repairs, and continuously lowering the average vibration levels of equipment in the plant.

2

Predictive maintenance techniques: Part 2

Vibration basics

2.1 Spring-mass system: mass, stiffness, damping

A basic understanding of how a discrete spring-mass system responds to an external force can be helpful in understanding, recognising and solving many problems encountered in vibration measurement and analysis.

Figure 2.1 shows a spring-mass system. There is a mass M attached to a spring with a stiffness k. The front of the mass M is attached to a piston with a small opening in it. The piston slides through a housing filled with oil.

The holed piston sliding through an oil-filled housing is referred to as a dashpot mechanism and it is similar in principle to shock absorbers in cars.

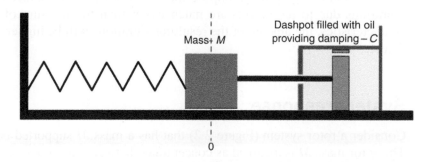

Figure 2.1
Spring-mass system

When an external force F moves the mass M forward, two things happen:

1. The spring is stretched.
2. The oil from the front of the piston moves to the back through the small opening.

We can easily visualize that the force F has to overcome three things:

1. Inertia of the mass M.
2. Stiffness of the spring k.
3. Resistance due to forced flow of oil from the front to the back of the piston or, in other words, the damping C of the dashpot mechanism.

All machines have the three fundamental properties that combine to determine how the machine will react to the forces that cause vibrations, just like the spring-mass system.

The three fundamental properties are:

(a) Mass (M)
(b) Stiffness (k)
(c) Damping (C).

These properties are the inherent characteristics of a machine or structure with which it will resist or oppose vibration.

(a) *Mass:*   Mass represents the inertia of a body to remain in its original state of rest or motion. A force tries to bring about a change in this state of rest or motion, which is resisted by the mass. It is measured in kg.
(b) *Stiffness:*   There is a certain force required to bend or deflect a structure with a certain distance. This measure of the force required to obtain a certain deflection is called stiffness. It is measured in N/m.
(c) *Damping:*   Once a force sets a part or structure into motion, the part or structure will have inherent mechanisms to slow down the motion (velocity). This characteristic to reduce the velocity of the motion is called damping. It is measured in N/(m/s).

As mentioned above, the combined effects to restrain the effect of forces due to mass, stiffness and damping determine how a system will respond to the given external force.

Simply put, a defect in a machine brings about a vibratory movement. The mass, stiffness and damping try to oppose the vibrations that are induced by the defect. If the vibrations due to the defects are much larger than the net sum of the three restraining characteristics, the amount of the resulting vibrations will be higher and the defect can be detected.

2.2   System response

Consider a rotor system (Figure 2.2) that has a mass M supported between two bearings. The rotor mass M is assumed as concentrated between the supported bearings; it contains an unbalance mass (Mu) located at a fixed radius r and is rotating at an angular velocity ω, where:

$$\omega = 2 \times \pi \times \frac{\text{rpm}}{60}$$

$$\text{rpm} = \text{revolutions per minute}$$

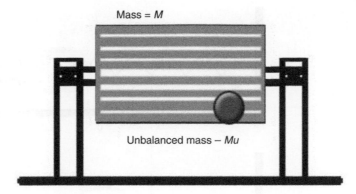

Mass = *M*

Unbalanced mass – *Mu*

Figure 2.2
A rotor system response

The vibration force produced by the unbalance mass *Mu* is represented by:

$$F(\text{unbalance}) = Mu \cdot r \cdot \omega^2 \cdot \sin(\omega t)$$

where t = time in seconds.

The restraining force generated by the three system characteristics is:

$$M \times (a) + C \times (v) + k \times (d)$$

where a = acceleration; v = velocity; d = displacement.

If the system is in equilibrium, the two forces are equal and the equation can be written as:

$$Mu \cdot r \cdot \omega^2 \cdot \sin(\omega t) = M \times (a) + C \times (v) + k \times (d)$$

However, in reality the restraining forces do not work in tandem. With changing conditions, one factor may increase while the other may decrease. The net result can display a variation in the sum of these forces.

This in turn varies the system's response (vibration levels) to exciting forces (defects like unbalance that generate vibrations). Thus, the vibration caused by the unbalance will be higher if the net sum of factors on the right-hand side of the equation is less than unbalance force. In a similar way, it is possible that one may not experience any vibrations at all if the net sum of the right-hand side factors becomes much larger than the unbalance force.

2.3   What is vibration?

Vibration, very simply put, is the motion of a machine or its part back and forth from its position of rest.

The most classical example is that of a body with mass *M* to which a spring with a stiffness *k* is attached. Until a force is applied to the mass *M* and causes it to move, there is no vibration.

Refer to Figure 2.3. By applying a force to the mass, the mass moves to the left, compressing the spring. When the mass is released, it moves back to its neutral position and then travels further right until the spring tension stops the mass. The mass then turns around and begins to travel leftwards again. It again crosses the neutral position and reaches the left limit. This motion can theoretically continue endlessly if there is no damping in the system and no external effects (such as friction).

This motion is called vibration.

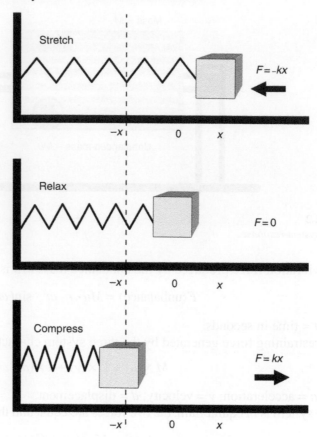

Figure 2.3
The nature of vibration

2.4   The nature of vibration

A lot can be learned about a machine's condition and possible mechanical problems by noting its vibration characteristics. We can now learn the characteristics, which characterize a vibration signal.

Referring back to the mass-spring body, we can study the characteristics of vibration by plotting the movement of the mass with respect to time. This plot is shown in Figure 2.4.

The motion of the mass from its neutral position, to the top limit of travel, back through its neutral position, to the bottom limit of travel and the return to its neutral position, represents one cycle of motion. This one cycle of motion contains all the information necessary to measure the vibration of this system. Continued motion of the mass will simply repeat the same cycle.

This motion is called periodic and harmonic, and the relationship between the displacement of the mass and time is expressed in the form of a sinusoidal equation:

$$X = X_0 \sin \omega t$$

X = displacement at any given instant t; X_0 = maximum displacement; $\omega = 2 \cdot \pi \cdot f$; f = frequency (cycles/s – hertz – Hz); t = time (seconds).

FREQUENCY = 0.25 cycles/s
(w) = 15 cycles/min (cpm)

PHASE	0	90	270	450	degrees

TIME	1	2	4	6	seconds

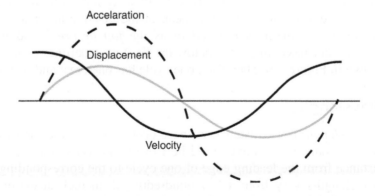

Figure 2.4
Simple harmonic wave – locus of spring-mass motion with respect to time

As the mass travels up and down, the velocity of the travel changes from zero to a maximum. Velocity can be obtained by time differentiating the displacement equation:

$$\text{velocity} = \frac{dX}{dt} = X_0 \cdot \omega \cdot \cos \omega t$$

Similarly, the acceleration of the mass also varies and can be obtained by differentiating the velocity equation:

$$\text{acceleration} = \frac{d(\text{velocity})}{dt} = - X_0 \cdot \omega^2 \cdot \sin \omega t$$

In Figure 2.5: displacement is shown as a *sine curve*; velocity, as a *cosine curve*; acceleration is again represented by a *sine curve*.

Accelaration

Displacement

Velocity

Figure 2.5
Waveform of acceleration, velocity and displacement of mass in simple harmonic motion

2.4.1 Wave fundamentals

Terms such as cycle, frequency, wavelength, amplitude and phase are frequently used when describing waveforms. We will now discuss these terms and others in detail as they are also used to describe vibration wave propagation.

We will also discuss waveforms, harmonics, Fourier transforms and overall vibration values, as these are concepts connected to machine diagnostics using vibration analysis.

In Figure 2.6, waves 1 and 2 have equal frequencies and wavelengths but different amplitudes. The reference line (line of zero displacement) is the position at which a particle of matter would have been if it were not disturbed by the wave motion.

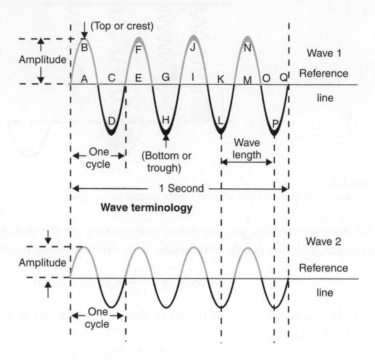

Figure 2.6
Comparison of waves with different amplitudes

2.4.2   Frequency (cycle)

At point E, the wave begins to repeat with a second cycle, which is completed at point I, a third cycle at point M, etc. The peak of the positive alternation (maximum value above the line) is sometimes referred to as the *top or crest*, and the peak of the negative alternation (maximum value below the line) is sometimes called the *bottom or trough*, as shown in Figure 2.6. Therefore, one cycle has one crest and one trough.

2.4.3   Wavelength

A *wavelength* is the distance in space occupied by one cycle of a transverse wave at any given instant. If the wave could be frozen and measured, the wavelength would be the distance from the leading edge of one cycle to the corresponding point on the next cycle. Wavelengths vary from a few hundredths of an inch at extremely high frequencies to many miles at extremely low frequencies, depending on the medium. In Figure 2.6 (wave 1), the distance between A and E, or B and F, etc., is one wavelength. The Greek letter λ (lambda) is commonly used to signify wavelength.

2.4.4   Amplitude

Two waves may have the same wavelength, but the crest of one may rise higher above the reference line than the crest of the other, for instance waves 1 and 2 in Figure 2.6. The height of a wave crest above the reference line is called the *amplitude* of the wave. The amplitude of a wave gives a relative indication of the amount of energy the wave transmits. A continuous series of waves, such as A through Q, having the same amplitude and wavelength, is called a train of waves or *wave train*.

2.4.5 Frequency and time

When a wave train passes through a medium, a certain number of individual waves pass a given point for a specific unit of time. For example, if a cork on a water wave rises and falls once every second, the wave makes one complete up-and-down vibration every second. The number of vibrations, or cycles, of a wave train in a unit of time is called the *frequency* of the wave train and is measured in *hertz* (Hz). If five waves pass a point in one second, the frequency of the wave train is five cycles per second. In Figure 2.6, the frequency of both waves 1 and 2 is four cycles per second (cycles per second is abbreviated as cps).

   In 1967, in honor of the German physicist Heinrich hertz, the term hertz was designated for use in lieu of the term 'cycle per second' when referring to the frequency of radio waves. It may seem confusing that in one place the term 'cycle' is used to designate the positive and negative alternations of a wave, but in another instance the term 'hertz' is used to designate what appears to be the same thing. The key is the time factor. The term cycle refers to any sequence of events, such as the positive and negative alternations, comprising one cycle of any wave. The term hertz refers to the number of occurrences that take place in one second.

2.4.6 Phase

If we consider the two waves as depicted in Figure 2.7, we find that the waves are identical in amplitude and frequency but a distance of *T*/4 offsets the crests of the waves. This lag of time is called the phase lag and is measured by the phase angle.

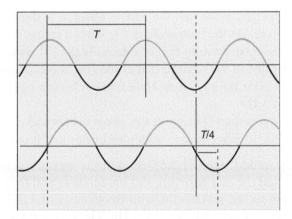

Figure 2.7
Phase relationship between two similar waves

   A time lag of *T* is a phase angle of 360°, thus a time lag of *T*/4 will be a phase angle of 90°.
   In this case we would normally describe the two waves as *out of phase* by 90°.

2.4.7 Waveforms

We have seen earlier, under the topic nature of vibrations, that displacement, velocity and acceleration of a spring-mass system in motion can be represented by sine and cosine waves. The waveform is a visual representation (or graph) of the instantaneous value of the motion plotted against time.

2.5 Harmonics

Figure 2.8 depicts many interesting waveforms. Let us presume that displacement is represented on the *Y*-axis. Since it is a representation vs time, the *X*-axis will be the time scale of 1 s.

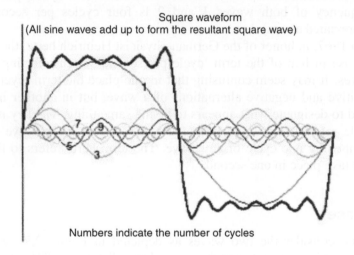

Square waveform
(All sine waves add up to form the resultant square wave)

Numbers indicate the number of cycles

Figure 2.8
An interesting waveform

- The first wave that we should observe is the [1] wave. It is represented by one cycle. As the time scale is 1 s, it has a frequency of 1 Hz.
- The next wave to be considered is the [3] wave. It can be seen that it has three cycles in the same period of the first wave. Thus, it has a frequency of 3 Hz.
- Third is the [5] wave. Here five cycles can be traced, and it thus has a frequency of 5 Hz.
- Next is the [7] wave. It has seven cycles and therefore a frequency of 7 Hz.
- The [9] wave is next with nine cycles and it will have a frequency of 9 Hz.

In this way an odd series (1,3,5,7,9...) of the waves can be observed in the figure. Such a series is called the *odd harmonics* of the fundamental frequency.

If we were to see waveforms with frequencies of 1,2,3,4,5 . . . Hz, then they would be the *harmonics* of the first wave of 1 Hz. The first wave of the series is usually designated as the wave with the *fundamental frequency*.

Coming back to the figure, it is noticed that if the fundamental waveforms with odd harmonics are added up, the resultant wave seen on the figure incidentally looks like a square waveform, which is more complex.

If a series of sinusoidal waveforms can be added to form a complex waveform, then is the reverse possible? It is possible and this is a widely used technique called the Fourier transform. It is a mathematically rigorous operation, which transforms waveforms from the time domain to the frequency domain and vice versa.

2.5.1   Fourier analysis

Fourier analysis is another term for the transformation of a time waveform (Figure 2.9) into a spectrum of amplitude vs frequency values. Fourier analysis is sometimes referred to as spectrum analysis, and can be done with a fast Fourier transform (FFT) analyzer.

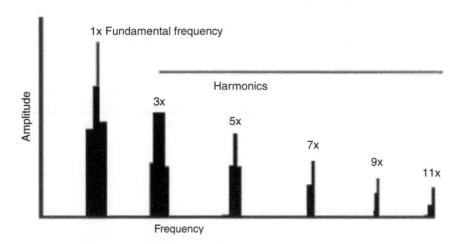

Figure 2.9
A Fourier transform of the square waveform

2.5.2   Overall amplitude

We have seen how a square waveform looks like in the time domain. The waveform is a representation of instantaneous amplitude of displacement, velocity or acceleration with respect to time.

The overall level of vibration of a machine is a measure of the total vibration amplitude over a wide range of frequencies, and can be expressed in acceleration, velocity or displacement (Figure 2.10).

The overall vibration level can be measured with an analog vibration meter, or it can be calculated from the vibration spectrum by adding all the amplitude values from the spectrum over a certain frequency range.

When comparing overall vibration levels, it is important to make sure they were calculated over the same frequency range.

2.5.3   Vibration terminology

Vibration displacement (peak to peak)

The total distance travelled by a vibrating part, from one extreme limit of travel to the other extreme limit of travel is referred to as the 'peak to peak' displacement.

- In SI units this is usually measured in 'microns' (1/1000th of a millimeter).
- In imperial units it is measured in 'mils' (milli inches – 1/1000th of an inch).

Displacement is sometimes referred to only as 'peak' (ISO 2372), which is half of 'peak to peak'.

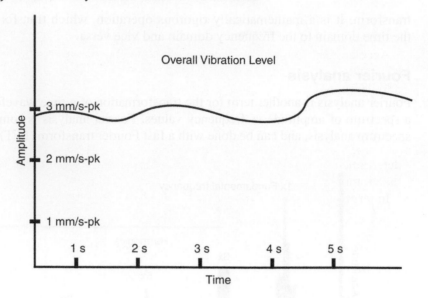

Figure 2.10
Overall vibration plot of velocity

Vibration velocity (peak)

As the vibrating mass moves, the velocity changes. It is zero at the top and bottom limits of motion when it comes to a rest before it changes its direction. The velocity is at its maximum when the mass passes through its neutral position. This maximum velocity is called as vibration velocity peak.

It is measured in mm/s-pk or inches/s-pk (ips-pk).

Vibration velocity (rms)

The International Standards Organization (ISO), who establishes internationally acceptable units for measurement of machinery vibration, suggested the velocity – root mean square (rms) as the standard unit of measurement. This was decided in an attempt to derive criteria that would determine an effective value for the varying function of velocity.

Velocity – rms tends to provide the energy content in the vibration signal, whereas the velocity peak correlated better with the intensity of vibration. Higher velocity – rms is generally more damaging than a similar magnitude of velocity peak.

Crest factor   The crest factor of a waveform is the ratio of the peak value of the waveform to the rms value of the waveform. It is also sometimes called the 'peak-to-rms-ratio'. The crest factor of a sine wave is 1.414, i.e. the peak value is 1.414 times the rms value. The crest factor is one of the important features that can be used to trend machine condition.

Vibration acceleration (peak)

In discussing vibration velocity, it was pointed out that the velocity of the mass approaches zero at extreme limits of travel. Each time it comes to a stop at the limit of travel, it must accelerate to increase velocity to travel to the opposite limit. Acceleration is defined as the rate of change in velocity.

Referring to the spring-mass body, acceleration of the mass is at a maximum at the extreme limit of travel where velocity of the mass is zero. As the velocity approaches a maximum value, the acceleration drops to zero and again continues to rise to its maximum value at the other extreme limit of travel.

Acceleration is normally expressed in g, which is the acceleration produced by the force of gravity at the surface of the earth. The value of g is 9.80665 m/s^2, 32.1739 ft/s^2 or 386.087 in./s^2.

Displacement, velocity, acceleration – which should be used?

The displacement, velocity and acceleration characteristics of vibration are measured to determine the severity of the vibration and these are often referred to as the 'amplitude' of the vibration.

In terms of the operation of the machine, the vibration amplitude is the first indicator to indicate how good or bad the condition of the machine may be. Generally, greater vibration amplitudes correspond to higher levels of machinery defects.

Since the vibration amplitude can be either displacement, velocity or acceleration, the obvious question is, which parameter should be used to monitor the machine condition?

The relationship between acceleration, velocity and displacement with respect to vibration amplitude and machinery health redefines the measurement and data analysis techniques that should be used. Motion below 10 Hz (600 cpm) produces very little vibration in terms of acceleration, moderate vibration in terms of velocity and relatively large vibrations in terms of displacement (see Figure 2.11). Hence, *displacement* is used in this range.

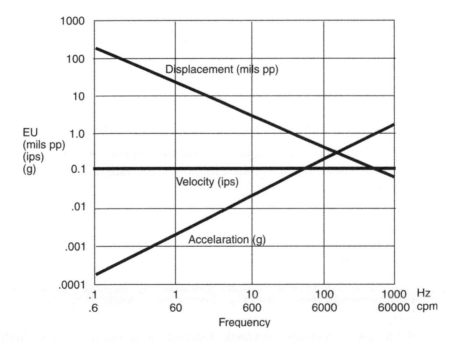

Figure 2.11
Relationship between displacement, velocity and acceleration at constant velocity. EU, engineering units

In the high frequency range, acceleration values yield more significant values than velocity or displacement. Hence, for frequencies over 1000 Hz (60 kcpm) or 1500 Hz (90 kcpm), the preferred measurement unit for vibration is *acceleration*.

It is generally accepted that between 10 Hz (600 cpm) and 1000 Hz (60 kcpm) *velocity* gives a good indication of the severity of vibration, and above 1000 Hz (60 kcpm), acceleration is the only good indicator.

Since the majority of general rotating machinery (and their defects) operate in the 10–1000 Hz range, velocity is commonly used for vibration measurement and analysis.

2.5.4   Using vibration theory to machinery fault detection

In Figure 2.12, a common machinery train is depicted. It consists of a driver or a prime mover, such as an electric motor. Other prime movers include diesel engines, gas engines, steam turbines and gas turbines. The driven equipment could be pumps, compressors, mixers, agitators, fans, blowers and others. At times when the driven equipment has to be driven at speeds other than the prime mover, a gearbox or a belt drive is used.

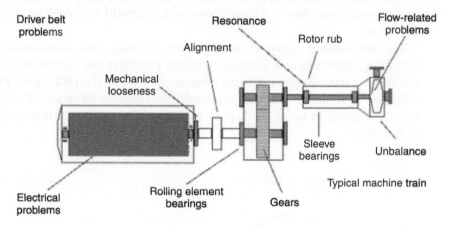

Figure 2.12
Machinery fault detection

Each of these rotating parts is further comprised of simple components such as:

- Stator (volutes, diaphragms, diffusers, stators poles)
- Rotors (impellers, rotors, lobes, screws, vanes, fans)
- Seals
- Bearings
- Couplings
- Gears
- Belts.

When these components operate continuously at high speeds, wear and failure is imminent. When defects develop in these components, they give rise to higher vibration levels.

With few exceptions, mechanical defects in a machine cause high vibration levels. Common defects that cause high vibrations levels in machines are:

- (a) Unbalance of rotating parts
- (b) Misalignment of couplings and bearings
- (c) Bent shafts
- (d) Worn or damaged gears and bearings
- (e) Bad drive belts and chains

    (f) Torque variations
    (g) Electromagnetic forces
    (h) Aerodynamic forces
    (i) Hydraulic forces
    (j) Looseness
    (k) Rubbing
    (l) Resonance.

To generalize the above list, it can be stated that whenever either one or more parts are unbalanced, misaligned, loose, eccentric, out of tolerance dimensionally, damaged or reacting to some external force, higher vibration levels will occur.

Some of the common defects are shown in Figure 2.12. The vibrations caused by the defects occur at specific vibration frequencies, which are characteristic of the components, their operation, assembly and wear. The vibration amplitudes at particular frequencies are indicative of the severity of the defects.

Vibration analysis aims to correlate the vibration response of the system with specific defects that occur in the machinery, its components, trains or even in mechanical structures.

2.6   Limits and standards of vibration

As mentioned above, vibration amplitude (displacement, velocity or acceleration) is a measure of the severity of the defect in a machine. A common dilemma for vibration analysts is to determine whether the vibrations are acceptable to allow further operation of the machine in a safe manner.

To solve this dilemma, it is important to keep in mind that the objective should be to implement regular vibration checks to detect defects at an early stage. The goal is not to determine how much vibration a machine will withstand before failure! The aim should be to obtain a trend in vibration characteristics that can warn of impending trouble, so it can be reacted upon before failure occurs.

Absolute vibration tolerances or limits for any given machine are not possible. That is, it is impossible to *fix* a vibration limit that will result in immediate machine failure when exceeded. The developments of mechanical failures are far too complex to establish such limits.

However, it would be also impossible to effectively utilize vibrations as an indicator of machinery condition unless some guidelines are available, and the experiences of those familiar with machinery vibrations have provided us with some realistic guidelines.

We have mentioned earlier that *velocity* is the most common parameter for vibration analysis, as most machines and their defects generate vibrations in the frequencies range of 10 Hz (600 cpm) to 1 kHz (60 kcpm).

2.6.1   ISO 2372

The most widely used standard as an indicator of vibration severity is ISO 2372 (BS 4675). The standard can be used to determine acceptable vibration levels for various classes of machinery. Thus, to use this ISO standard, it is necessary to first classify the machine of interest. Reading across the chart we can correlate the severity of the machine condition with vibration. The standard uses the parameter of velocity-rms to indicate severity. The letters A, B, C and D as seen in Figure 2.13, classify the severity.

ISO 2372 – ISO Guideline for Machinery Vibration Severity						
Ranges of Vibration severity		Examples of quality judgment for separate classes of machines				
Velocity – in/s – Peak	Velocity – mm/s – rms	Class I	Class II	Class III	Class IV
0.015	0.28				
0.025	0.45				
0.039	0.71				
0.062	1.12				
0.099	1.8				
0.154	2.8				
0.248	4.5				
0.392	7.1				
0.617	11.2				
0.993	18				
1.54	28				
2.48	45				
3.94	71				

A – Good
B – Acceptable
C – Still acceptable
D – Not acceptable

Figure 2.13
ISO 2372 – ISO guideline for machinery vibration severity

Class I Individual parts of engines and machines integrally connected with a complete machine in its normal operating condition (production electrical motors of up to 15 kW are typical examples of machines in this category).

Class II Medium-sized machines (typically electrical motors with 15–75 kW output) without special foundations, rigidly mounted engines or machines (up to 300 kW) on special foundations.

Class III Large prime movers and other large machines with rotating masses mounted on rigid and heavy foundations, which are relatively stiff in the direction of vibration.

Class IV Large prime movers and other large machines with rotating masses mounted on foundations, which are relatively soft in the direction of vibration measurement (for example – turbogenerator sets, especially those with lightweight substructures).

American Petroleum Institute (API specification)

The American Petroleum Institute (API) has set forth a number of specifications dealing with turbomachines used in the petroleum industry. Some of the specifications that have been prepared include API-610, API-611, API-612, API-613, API-616 and API-617. These specifications mainly deal with the many aspects of machinery design, installation, performance and support systems. However, there are also specifications for rotor balance quality, rotor dynamics and vibration tolerances.

API standards have developed limits for casing as well as shaft vibrations (Figure 2.14).

The API specification on vibration limits for turbo machines is widely accepted and followed with apparently good results.

The API standard specifies that the maximum allowable vibration displacement of a shaft measured in mils (milli-inches = 0.001 inch = 0.0254 mm) peak–peak shall not be greater than 2.0 mils or $(12\,000/N)^{1/2}$, where N is speed of the machine, whichever is less.

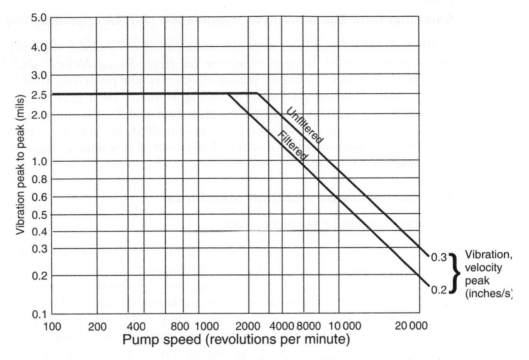

Bearing – housing vibration limits (anti-friction bearings)

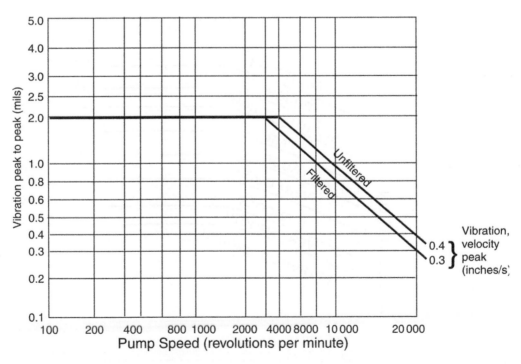

Shaft vibration limits (sleeve bearings)

Figure 2.14
Vibration limits – API-610 centrifugal pumps in refinery service

American Gear Manufacturers Association (AGMA specification)

In 1972, AGMA formulated a specification called the AGMA standard specification for *Measurement of Lateral Vibration on High Speed Helical and Herringbone Gear Units – AGMA 426.01* (the present standard is now revised to AGMA 6000-B96).

It presents a method for measuring linear vibration on a gear unit. It recommends instrumentation, measuring methods, test procedures and discrete frequency vibration limits for acceptance testing. It annexes a list of system effects on gear unit vibration and system responsibility. Determination of mechanical vibrations of gear units during acceptance testing is also mentioned.

2.6.2 IRD mechanalysis vibration standards

General machinery severity chart

The general machinery severity chart (Figure 2.15) incorporates vibration velocity measurements along with the familiar displacement measurements, when amplitude readings are obtained in metric units (microns-peak–peak or mm/s-peak). The chart evolved out of a large amount of data collected from different machines.

When using displacement measurements, only filtered displacement readings (for a specific frequency) should be applied to the chart. Overall vibration velocity can be applied since the lines that divide the severity regions are actually constant velocity lines. The chart is used for casing vibrations and not meant for shaft vibrations.

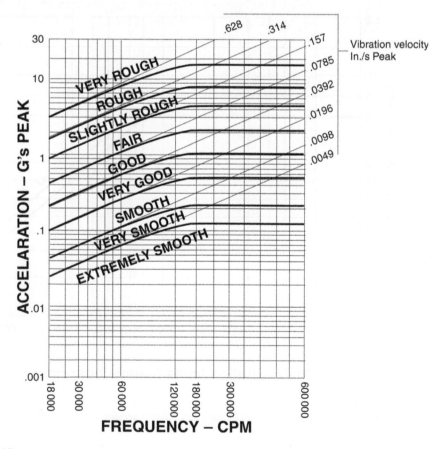

Figure 2.15
General machinery severity chart

The chart applies to machines that are rigidly mounted or bolted to a fairly rigid foundation. Machines mounted on resilient vibration isolators such as coil springs or rubber pads will generally have higher amplitudes of vibration compared to rigidly mounted machines.

A general rule is to allow twice as much vibration for a machine mounted on isolators. High-frequency vibrations should not be subjected to the above criteria.

General vibration acceleration severity chart

The general vibration acceleration severity chart is used in cases where machinery vibration is measured in units of acceleration (g-peak) (see Figure 2.16).

Constant vibration velocity lines are included on the chart to provide a basis for comparison, and it can be noted that for vibration frequencies below 60 000 cpm (1000 Hz), the lines that divide the severity regions are of a relatively constant velocity. However, above this limit, the severity regions are defined by nearly constant acceleration values.

Since the severity of vibration acceleration depends on frequency, only filtered acceleration readings can be applied to the chart.

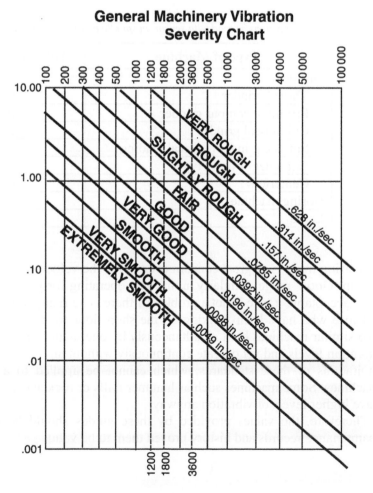

Figure 2.16
Vibration acceleration severity chart – IRD mechanalysis

Tentative guide to vibration limits for machine tools

Amplitudes of machine tool vibration must be relatively low in order to maintain dimensional tolerances and to provide acceptable surface finish of machined workpieces.

The vibration limits tabulated below are based on the experience of manufacturers and were selected as typical of those required on machine tools in order to achieve these objectives.

These limits should be used as a guide only – modern machines may need even tighter limits for stringent machining specifications.

It should be mentioned that vibration limits are in displacement units, as the primary concern for machine tool vibration is the relative motion between the workpiece and the cutting edge. This relative motion is compared to the specified surface finish and dimensional tolerances, which are also expressed in terms of displacement units.

When critical machinery with a heavy penalty for process downtime is involved, the decision to correct a condition of vibration is often a very difficult one to make. Therefore, when establishing acceptable levels of machinery condition, experience and factors such as safety, labor costs, downtime costs and the machine's criticality should be considered.

It is thus reiterated that standards should only be an indicator of machine condition and not a basis for shutting down the machine. What is of extreme importance is that vibrations of machines should be recorded and trended diligently.

Displacement of vibrations as read with sensor on spindle bearing housing in the direction of cut

Type of Machine	Tolerance Range (mils)
Grinders	
Thread grinder	0.01–0.06
Profile or contour grinder	0.03–0.08
Cylindrical grinder	0.03–0.10
Surface grinder (vertical reading)	0.03–0.2
Gardener or besly type	0.05–0.2
Centerless	0.04–0.1
Boring machine	0.06–0.1
Lathe	0.2–1

A rising trend is of great concern even when the velocity values as per the standard are still in 'Good' range. Similarly, a machine operating for years with velocity values in the 'Not acceptable' range is not a problem if there is no rising trend.

Those who have been working on the shop floor for a long time will agree that even two similar machines built simultaneously by one manufacturer can have vastly different vibration levels and yet operate continuously without any problems. One has to accept the limitations of these standards, which cannot be applied to a wide range of complex machines. Some machines such as hammer mills or rock and coal crushers will inherently have higher levels of vibration anyway.

Therefore, the values provided by these guides should be used only if experience, maintenance records and history proved them to be valid.

3

Data acquisition

3.1   Introduction

The topics discussed in the previous sections were theoretical in nature, introducing the basics of vibration. With *data acquisition*, we take the first steps into the domain of practical vibration analysis. It includes the following main tasks:

- Collection of machinery vibration
- Conversion of the vibration signal to an electrical signal
- Transformation of the electrical signal to its components
- Providing information and documentation related to vibration data.

The above entails the entire hardware of the vibration analysis system or program. It includes transducers, electronic instruments that store and analyze data, the software that assist in vibration analysis, record keeping and documentation.

3.2   Collection of vibration signal – vibration transducers, characteristics and mountings

To measure machinery or structural vibration, a transducer or a vibration pickup is used. A transducer is a device that converts one type of energy, such as vibration, into a different type of energy, usually an electric current or voltage.

Commonly used transducers are velocity pickups, accelerometers and Eddy current or proximity probes. Each type of transducer has distinct advantages for certain applications, but they all have limitations as well. No single transducer satisfies all measurement needs. One of the most important considerations for any application is to select the transducer that is best suited for the job.

The various vibration transducers are discussed below.

3.2.1   Velocity pickup

The velocity pickup is a very common transducer for monitoring the vibration of rotating machinery. This type of vibration transducer installs easily on most analyzers, and is rather inexpensive compared to other sensors. For these reasons, the velocity transducer is ideal for general purpose machine-monitoring applications. Velocity pickups have been used as vibration transducers on rotating machines for a very long time, and these are still utilized for a variety of applications today. Velocity pickups are available in many different physical configurations and output sensitivities.

3.2.2   Theory of operation

When a coil of wire is moved through a magnetic field (coil-in-magnet type) (Figure 3.1), a voltage is induced across the end wires of the coil. The transfer of energy from the flux field of the magnet to the wire coil generates the induced voltage. As the coil is forced through the magnetic field by vibratory motion, a voltage signal correlating with the vibration is produced.

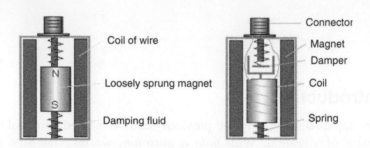

Figure 3.1
Two basic types of velocity pickups employing principle of motion of magnet-in-coil and coil-in-magnet

The magnet-in-coil type of sensor is made up of three components: a permanent magnet, a coil of wire and spring supports for the magnet. The pickup is filled with oil to dampen the spring action. The relative motion between the magnet and coil caused by the vibration motion induces a voltage signal.

The velocity pickup is a self-generating sensor and requires no external devices to produce a voltage signal. The voltage generated by the pickup is directly proportional to the velocity of the relative motion.

Due to gravity forces, velocity transducers are manufactured differently for horizontal or vertical axis mounting. The velocity sensor has a sensitive axis that must be considered when applying them to rotating machinery. Velocity sensors are also susceptible to cross-axis vibration, which could damage a velocity sensor.

Wire is wound on a hollow bobbin to form the wire coil. Sometimes, the wire coil is counter wound (wound in one direction and then in the opposite direction) to counteract external electrical fields. The bobbin is supported by thin, flat springs to position it accurately in the permanent magnet's field.

A velocity signal produced by vibratory motion is normally sinusoidal in nature. Thus, in one cycle of vibration, the signal reaches a maximum value twice. The second maximum value is equal in magnitude to the first maximum value, but opposite in direction.

Another convention to consider is that motion towards the bottom of a velocity transducer will generate a positive output signal. In other words, if the transducer is held in its sensitive axis and the base is tapped, the output signal will be positive when it is initially tapped.

3.2.3   Number of sensors

All vibration sensors measure motion along their major axis. This fact should be taken into account when choosing the number of sensors to be used. Due to the structural asymmetry of machine cases, the vibration signals in the vertical, horizontal and axial directions (with respect to the shaft) may differ. Where possible, a velocity transducer should be mounted in the vertical, horizontal and axial planes to measure vibration in the three directions. The three sensors will provide a complete picture of the vibration signature of the machine.

Mounting

For the best results, the mounting location must be flat, clean and slightly larger than the velocity pickup. If it is possible, it should be clamped with a separate mounting enclosure. The surface will have to be drilled and tapped to accommodate the mounting screw of the sensor. Whenever a velocity pickup is exposed to hazardous environments such as high temperatures, radioactivity, water or magnetic fields, special protection measures should be taken.

Magnetic interferences should especially be taken into account when measuring vibrations of large AC motors and generators. The alternating magnetic field that these machines produce may affect the coil conductor by inducing a voltage in the pickup that could be confused with actual vibration. In order to reduce the effect of the alternating magnetic field, magnetic shields can be used.

A quick method to determine whether a magnetic shield would be required is to hang the pickup close to the area where vibrations must be taken (with a steady hand as not to induce real vibrations). If significant vibrations are observed, a magnetic shield may be required.

Sensitivity

Some velocity pickups have the highest output sensitivities of all the vibration pickups available for rotating machine applications. Higher output sensitivity is useful in situations where induced electrical noise is a problem. Larger sensor outputs for given vibration levels will be influenced less by electrical noise.

Sensitivities are normally expressed in mV/in./s or mV/mm/s. General values are in the range of 500 mV/in./s to 750 mV/in./s (20–30 mV/mm/s). The sensitivity of the velocity pickup is constant over a specified frequency range, usually between 10 Hz and 1 kHz. At low frequencies of vibration, the sensitivity decreases because the pickup coil is no longer stationary with respect to the magnet, or vice versa. This decrease in pickup sensitivity usually starts at a frequency of approximately 10 Hz, below which the pickup output drops exponentially. The significance of this fact is that amplitude readings taken at frequencies below 10 Hz using a velocity pickup are inaccurate.

Even though the sensitivity may fall at lower frequencies this does not prevent the usage of this pickup for repeated vibration measurement at the same position only for trending or balancing.

Frequency response

Velocity pickups have different frequency responses depending on the manufacturer. However, most pickups have a linear frequency response range in the order of 10 Hz–1 kHz. This is an important consideration when selecting a velocity pickup for a rotating machine application. The pickup's frequency response must be within the expected frequency range of the machine.

Calibration

Velocity pickups should be calibrated on an annual basis. The sensor should be removed from service for calibration verification. Verification is necessary because velocity pickups are the only industrial vibration sensors with internal moving parts that are subject to fatigue failure.

This verification should include a sensitivity response vs frequency test. This test will determine if the internal springs and damping system have degraded due to heat and vibration. The test should be conducted with a shaker capable of variable amplitude and frequency testing.

Advantages

- Ease of installation
- Strong signals in mid-frequency range
- No external power required.

Disadvantages

- Relatively large and heavy
- Sensitive to input frequency
- Narrow frequency response
- Moving parts
- Sensitive to magnetic fields.

Acceleration transducers/pickup

Accelerometers are the most popular transducers used for rotating machinery applications (Figure 3.2). They are rugged, compact, lightweight transducers with a wide frequency response range. Accelerometers are extensively used in many condition-monitoring applications. Components such as rolling element bearings or gear sets generate high-vibration frequencies when defective. Machines with these components should be monitored with accelerometers.

The installation of an accelerometer must carefully be considered for an accurate and reliable measurement.

Accelerometers are designed for mounting on machine cases. This can provide continuous or periodic sensing of absolute case motion (vibration relative to free space) in terms of acceleration.

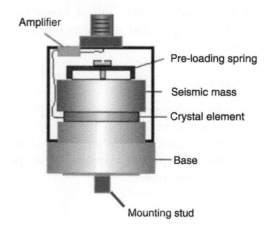

Figure 3.2
Accelerometer

Theory of operation

Accelerometers are inertial measurement devices that convert mechanical motion into a voltage signal. The signal is proportional to the vibration's acceleration using the piezoelectric principle. Inertial measurement devices measure motion relative to a mass. This follows Newton's third law of motion: body acting on another will result in an equal and opposite reaction on the first.

Accelerometers consist of a piezoelectric crystal (made of ferroelectric materials like lead zirconate titanate and barium titanate) and a small mass normally enclosed in a protective metal case.

When the accelerometer is subjected to vibration, the mass exerts a varying force on the piezoelectric crystal, which is directly proportional to the vibratory acceleration. The charge produced by the piezoelectric crystal is proportional to the varying vibratory force.

The charge output is measured in Pico-coulombs per g (pC/g) where g is the gravitational acceleration. Some sensors have an internal charge amplifier, while others have an external charge amplifier. The charge amplifier converts the charged output of the crystal to a proportional voltage output in mV/g.

Current or voltage mode

This type of accelerometer has an internal, low-output impedance amplifier and requires an external power source. The external power source can be either a constant current source or a regulated voltage source. This type of accelerometer is normally a two-wire transducer with one wire for the power and signal, and the second wire for common. They have a lower-temperature rating due to the internal amplifier circuitry. Output cable lengths up to 500 feet have a negligible effect on the signal quality. Longer cable lengths will reduce the effective frequency response range.

Charge mode

Charge mode accelerometers differ slightly from current or voltage mode types. These sensors have no internal amplifier and therefore have a higher-temperature rating. An external charge amplifier is supplied with a special adaptor cable, which is matched to the accelerometer. Field wiring is terminated to the external charge amplifier. As with current or voltage mode accelerometers, signal cable lengths up to 500 feet have negligible effect on the output signal quality. Longer cable lengths will reduce the effective frequency response range.

Mounting

It is important to know the possible mounting methods for this vibration sensor. Four primary methods are used for attaching sensors to monitoring locations. These are *stud* mounted, *adhesive* mounted, *magnet* (double leg or flat) mounted and non-mounted – e.g. using *handheld probes or stingers*. Each method affects the high-frequency response of the accelerometer. Stud mounting provides the widest frequency response and the most secure, reliable attachment.

The other three methods reduce the upper frequency range of the sensor. In these cases, the sensor does not have a very secure direct contact with the measurement point. Inserting mounting pieces, such as adhesive pads, magnets or probe tips, introduces a mounted resonance. This mounted resonance is lower than the natural resonance of the sensor and reduces the upper frequency range. A large mounting piece causes lower mounted resonance and also lowers the usable frequency range of the transducer.

The mounting methods typically used for monitoring applications are discussed in more detail below.

The *stud/bolt mounting method* is the best method available for permanent mounting applications. This method is accomplished by screwing the sensor in a stud or a machined block. This method permits the transducer to measure vibration in the most ideal manner and should be used wherever possible.

The mounting location for the accelerometer should be clean and paint-free. The mounting surface should be spot-faced to achieve a smooth surface. The spot-faced diameter should be slightly larger than the accelerometer diameter. Any irregularities in the mounting surface preparation will translate into improper measurements or damage to the accelerometer.

The *adhesive or glue mounting* method provides a secure attachment without extensive machining. However, when the accelerometer is glued, it typically reduces the operational frequency response range or the accuracy of the measurement. This reduction is due to the damping qualities of the adhesive. Also, replacement or removal of the accelerometer is more difficult than with any other attachment method. For proper adhesive bonding, surface cleanliness is of extreme importance.

The *magnetic mounting* method is typically used for temporary measurements with a portable data collector or analyzer. This method is not recommended for permanent monitoring. The transducer may be inadvertently moved and the multiple surfaces and materials of the magnet may interfere with high-frequency signals.

By design, accelerometers have a natural resonance which is 3–5 times higher than the high end of the rated frequency response. The frequency response range is limited in order to provide a flat response over a given range. The rated range is achievable only through stud mounting. As mentioned before, any other mounting method adversely affects the resonance of the sensor, such as the reliable usable frequency range.

Sensitivity

Accelerometers utilized for vibration monitoring are usually designed with a sensitivity of 100 mV/g. Other types of accelerometers with a wide range of sensitivities for special applications such as structural analysis, geophysical measurement, very high frequency analysis or very low speed machines are also available.

Frequency range

Accelerometers are designed to measure vibration over a given frequency range. Once the particular frequency range of interest for a machine is known, an accelerometer can be selected. Typically, an accelerometer for measuring machine vibrations will have a frequency range from 1 or 2 Hz to 8 or 10 kHz. Accelerometers with higher-frequency ranges are also available.

Calibration

Piezoelectric accelerometers cannot be recalibrated or adjusted. Unlike a velocity pickup, this transducer has no moving parts subject to fatigue. Therefore, the output sensitivity does not require periodic adjustments. However, high temperatures and shock can damage the internal components of an accelerometer.

When the reliability of an accelerometer is doubtful, a simple test of the transducer's bias voltage can be used to determine whether it should be removed from service.

An accelerometer's bias voltage is the DC component of the transducer's output signal. The bias voltage is measured with a DC voltmeter across the transducer's output and common leads with the power on. At the same time, the power supply should also be checked to eliminate the possibility of improper power voltage affecting the bias voltage level of the sensor.

3.2.4 Displacement probes – Eddy current transducers – proximity probes

Eddy current transducers (proximity probes) are the preferred vibration transducers for vibration monitoring on journal bearing equipped rotating machinery. Typical applications are predominantly high-speed turbomachinery.

Eddy current transducers are the only transducers that provide displacement of shaft or shaft-relative (shaft relative to the bearing) vibration measurements. Several methods are usually available for the installation of Eddy current transducers, including internal, internal/external, and external mounting.

Theory of operation

An Eddy current system is a matched component system which consists of a probe, an extension cable and an oscillator/demodulator (Figure 3.3).

A high-frequency radio frequency (RF) signal at 2 MHz is generated by the oscillator/demodulator. This is sent through the extension cable and radiated from the probe tip.

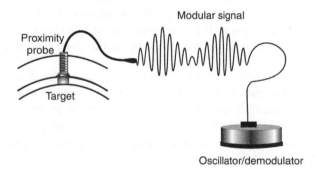

Figure 3.3
Proximity probe principle

Eddy currents are generated in the surface of the shaft. The oscillator/demodulator demodulates the signal and provides a modulated DC voltage, where the DC portion is directly proportional to the gap (distance) and the AC portion is directly proportional to vibration.

In this way, an Eddy current transducer can be used for both radial vibration and distance measurements such as the *axial thrust position* and *shaft position*.

Number of transducers

All vibration transducers measure motion in their mounted plane. In other words, shaft motion is either directed away from or towards the mounted Eddy current probe, and thus the radial vibration is measured in this way.

On smaller, less critical machines, one Eddy current transducer system per bearing is adequate. The single Eddy current probe measures the shaft's vibration in that given plane. Therefore, the Eddy current probe should be mounted in the plane where the largest vibrations are expected.

On larger, more critical machines, two Eddy current transducer systems are normally recommended per bearing. The probes for this type of installation are mounted 90° apart from each other. Since the probes will measure the vibration in their respective planes, the shaft's total movement within the journal bearing is measured. An 'Orbit' or Cartesian product of the two vibration signals can be constructed when both Eddy current transducers are connected to an oscilloscope.

Mounting methods

Orientation of transducer(s)   As most of the bearing housings on which probes are attached are horizontally split, transducers are commonly mounted at 45° on both sides of the vertical plane. If possible, the orientation of the transducers should be consistent along the length of the machine train for easier diagnostics. In all cases, the orientations should be well documented.

Perpendicular to shaft centreline   Care must be exercised in all installations to ensure that the Eddy current probes are mounted perpendicular to the shaft centreline. Deviation by more than 1–2° will affect the output sensitivity of the system.

Probe side clearances   The RF field emitted from the probe tip of an Eddy current transducer is shaped like a cone at approximately 45° angles. Clearance must be provided on all sides of the probe tip to prevent interference with the RF field. For instance, if a hole is drilled in a bearing for probe installation, it must be counter-bored to prevent side clearance interference. It is important to ensure that collars or shoulders on the shaft do not thermally grow under the probe tip as the shaft expands due to heat.

Internal mounting   During *internal mounting*, the Eddy current probes are mounted inside the machine or bearing housing with a special bracket (Figure 3.4). The transducer system is installed and gapped properly prior to the bearing cover being reinstalled. Provision must be made for the transducer's cable protruding from the bearing housing. This can be accomplished by using an existing plug or fitting, or by drilling and tapping a hole above the oil line.

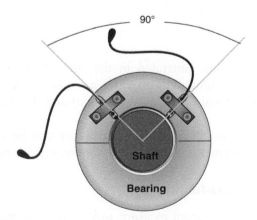

Figure 3.4
Internally mounted probes

The transducer's cables must also be tied down within the bearing housing to prevent cable failure from 'windage'. For added safety and reliability, all fasteners inside the bearing housing should be safety wired.

Advantages of internal mounting
   Less machining required for installation.
   True bearing-relative measurement is possible.
   The Eddy probe has an unconstrained view on the shaft surface.

Disadvantages of internal mounting
   There is no access to probe while the machine is running.
   Cables must be tied down with extreme care, because they might break due to 'windage'.
   Transducer cable exits must be provided.
   Care must be taken to avoid oil leakage.

External/internal mounting   External/internal mounting is accomplished when Eddy probes are mounted with a mounting adaptor (Figure 3.5). These adaptors allow external access to the probe, but the probe tip itself is inside the machine or bearing housing. While drilling and tapping the bearing housing or cover, it is important to ensure that the Eddy probes are installed perpendicular to the shaft centerline.

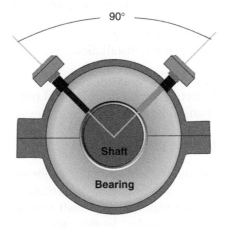

Figure 3.5
External/internal mounted probes

   In some cases, due to space limitations, external/internal mounting is accomplished by drilling or making use of existing holes in the bearing itself, usually at an oil-return groove.

Advantages of external/internal mounting
   Eddy probe replacement is possible while machine is running.
   Eddy probe has an unconstrained view on the shaft.
   Gap may be changed while machine is running.

Disadvantages of external/internal mounting
   May not be true bearing-relative measurement.
   More machining required.
   Long probe/stinger length may cause resonance.

External mounting   Most old turbomachines were not equipped with radial probes. The construction of older machines may not provide ideal installation of probes. External Eddy probes are mounted on such machines (Figure 3.6). It is usually a last resort installation.

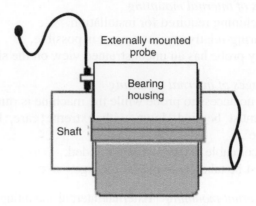

Figure 3.6
Externally mounted probes

The only valid reason for using this method is inadequate space available within the bearing housing for internal mounting. Special care must be given to the Eddy probe viewing area, and mechanical protection must be provided to the transducer and cable.

Advantages of external mounting
   It is the most inexpensive installation.

Disadvantages of external mounting
   It may record electrical and/or mechanical runout of the shaft.
   Requires mechanical protection.

Target material/target area

Eddy current transducers are normally calibrated for 4140 (Carbon Steel) steel unless otherwise specified. Because Eddy currents are sensitive to the permeability and resistivity of the shaft material, any shaft material other than 4000 series steels must be specified at the time of order. In the case of another kind of shaft material, the probe supplier might require a sample of the shaft material.

Mechanical runout

Eddy current transducers are also sensitive to the shaft smoothness for radial vibration. A smooth (ground/polished) area approximately three times the diameter of the probe must be provided for as a viewing area. The selected journal area on most shafts is wider than the bearing itself, allowing for probe installation directly adjacent to the bearing.

Electrical runout

Since Eddy current transducers are sensitive to the permeability and resistivity of the target material and also because the field of the transducer extends into the surface area of the shaft by approximately 0.4 mm (15 mils), care must be taken to avoid non-homogeneous viewing area materials such as chrome. Another form of electrical runout can be caused by small magnetic fields, such as those left by magna-fluxing without proper degaussing.

Calibration

All Eddy current systems (probe, cable and oscillator/demodulator) must be calibrated prior to commissioning. This is done using a static calibrator, –24 V DC power supply and a digital voltmeter. The probe is installed in the tester with the target set against the probe tip. The micrometer with the target attached is then rotated away from the probe in increments of 0.1 mm (or 5 mils). The voltage reading is recorded and plotted at each increment. The graph obtained for the specified range should be linear.

Probe to target gap

When installed, Eddy current probes must be gapped properly. In most radial vibration applications, adjusting the gap of the transducer to the center of the linear range is adequate. For example, as shown in Figure 3.7, a gap set for –12.0 V DC using a digital voltmeter would correspond to an approximate mechanical gap of 1.5 mm (or 60 mils).

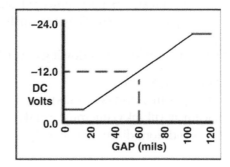

Figure 3.7
Typical calibration curve

The voltage method of gapping the probe is recommended above mechanical gapping. In all cases, final probe gap voltage should be documented and kept in a safe place.

Proximity probes typically have a sensitivity of 10 mV/µm (or 200 mV/mil) with an approximate linear range of 2–2.5 mm (or 80–100 mils).

Proximity probes are set at –8 to –9 V, which creates a clearance of 1.0–1.3 mm. This voltage setting will place the probe in the middle of its linear range, thus allowing the probe to sense movement in the positive direction and in the negative direction.

Proximity probes should be located at the bearing, and no further than 25 mm (or one inch) axially from the bearing towards the center of the shaft.

3.3 Conversion of vibrations to electrical signal

Vibration transducers convert the physical vibration motion of a machine into an electrical signal. However, this signal in the raw form is of no use unless it is processed to provide meaningful information that can be related to machine condition. Thus, there is a need for monitoring equipment that can take such an electrical signal from a transducer and process it into meaningful data.

Also, in the earlier topics, we have discussed the adoption of various maintenance philosophies applied in process plants or on the shop floor, based on the equipment classification. The type of monitoring methods to be used for each different machine is also based on the above rationale. Once the machinery monitoring needs are established, the next step is to select suitable monitoring equipment that fulfills these requirements the best.

The various options commercially available are:

- Handheld vibration meters and analyzers
- Portable data collectors/analyzers
- Vibration analysis/database management software
- Permanent online data acquisition and analysis equipment.

3.3.1   Handheld vibration meters and analyzers

A handheld vibration meter is an inexpensive and simple-to-use instrument that is an essential part of any vibration program. Plant operators and vibration technicians carry handheld meters and analyzers on their routine rounds.

When these are held in contact with machinery, they provide a display of vibration levels (either analog or digital). The readout provides immediate information that can be used to determine if the overall vibration levels are normal or abnormal. Handheld vibration meters are typically battery powered and use an accelerometer for sensing. Sometimes a velocity pickup is used. They are small, lightweight and rugged for day-to-day use (Figure 3.8).

Handheld meters can provide the following data (depending on the specific model):

- Acceleration (pk) (g)
- Velocity (pk-rms) (mm/s or in./s)
- Displacement (pk-pk) (microns or mils)
- Bearing condition (discussed later) (gSE, dB and others).

Advantages
They are convenient and flexible, and require very little skill to use. It is an inexpensive starting point for any new condition-monitoring program.

Disadvantages
Limited in the type of measurements that they can perform. They also lack data storage capability (however, some instruments are now available with some limited storage capacity).

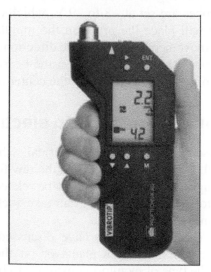

Figure 3.8
Handheld vibration meter from Vibrotip

Collecting and reporting vibration data – handheld meters

Handheld vibration measurements are quite common. This data acquisition method is rapid, convenient and demands minimal skills. The first step is to identify the positions on the machine from where measurements should be taken.

Mechanical vibrations have an analogy to an electrical current. Just as an electric current would tend to go to earth, vibrations caused by defects in rotating machinery would travel to the ground through its supports. It has been shown that bearings are the location where the vibrations 'jump' from the rotor to the stator to ultimately get grounded. Thus, it is at the bearings where the best signals for condition monitoring can be measured and hence these are generally the best positions for vibration measurements.

It is always necessary to follow a convention for labeling the various bearings of a machine train from where measurements were made. The general convention is to start labeling from where the power comes in. For example, a simple machine train consisting of a motor and pump will be labeled in the following manner (Figure 3.9):

- Motor non-drive end bearing – A
- Motor drive end bearing – B
- Pump outboard bearing (next to the coupling) – C
- Pump inboard bearing (away from coupling) – D.

Once the bearings are labeled, it is important that vibrations are taken in the three Cartesian directions. In vibration nomenclature, these are the vertical, horizontal and axial directions. This is necessary due to the construction of machines – their defects can show up in any of the three directions and hence each should be measured.

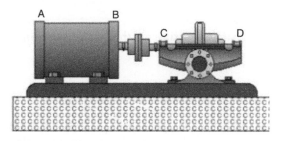

Figure 3.9
Bearing naming nomenclature

Vibration level reporting is generally done in the manner shown below.

Velocity – mm/s-pk (20/8/01)	Vertical	Horizontal	Axial
A	2.4	1.7	1.0
B	2.1	1.9	1.2
C	4.3	5.6	2.7
D	3.7	4.1	2.1

Handheld measurements are subject to a number of errors. It is thus important that personnel carrying out this task are aware of the possible errors that can occur while taking measurements. Errors can occur due to:

- Position on machinery
- Probe angle

- Probe type
- Mounting method of sensor (or pressure applied if handheld).

Position on machinery

Measurements should be taken at exactly the same location to enable direct comparisons of data sets. Moving the probe only a small distance on a machine can produce drastically different vibration levels. To ensure measurements are taken at the same spot, it should be marked with paint, or a shallow conical hole should be drilled for identification.

Probe angle

The sensor or the probe should always be oriented perpendicular to the machine surface. Tilting the probe slightly at an angle may induce an error.

Probe type

Some handheld meters are supplied with probes (called stingers) and also round magnets, which can be screwed into velocity transducers or accelerometers. Measuring vibration with magnetic attachments can collect higher vibration frequencies than what can be measured with handheld probes. When collecting vibration data on a machine generating high frequencies with a handheld meter, changing the probe type will show a drastic difference in the overall levels.

Pressure

Even and consistent pressure of the hand is required to get comparable readings with handheld meters.

3.3.2   Portable data collectors/analyzers

Modern data collectors/analyzers can provide information of any vibration characteristics in any desired engineering unit. There are basically two types of data collectors and analyzers (Figure 3.10):

- Single channel
- Dual channel.

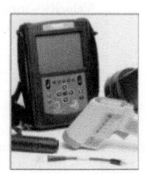

Figure 3.10
Commercially available data collectors/analyzers with accessories kit

Advantages

Can collect, record and display vibration data such as FFT spectra, overall trend plots and time domain waveforms.

Provides orderly collection of data.

Automatically reports measurements out of pre-established limit thresholds.

Can perform field vibration analysis.

Disadvantages

They are relatively expensive.

Operator must be trained for use.

Limited memory capability and thus data must be downloaded after collection.

3.3.3 Vibration analysis – database management software

The data collector/analyzer can collect and store only a limited amount of data. Therefore, the data must be downloaded to the computer to form a history and long-term machinery information database for comparison and trending. To perform the above tasks, and also aid in collection, management and analysis of machinery data, database management software packages are required.

These database management programs for machinery maintenance store vibration data and make comparisons between current measurements, past measurements and pre-defined alarm limits. Measurements transferred to the vibration analysis software are rapidly investigated for deviations from normal conditions. Overall vibration levels, FFTs, time waveforms and other parameters are produced to help analyze these vibration changes.

Reports can be generated showing machines whose vibration levels cross alarm thresholds. Current data are compared to baseline data for analysis and also trended to show vibration changes over a period of time. Trend plots give early warnings of possible defects and are used to schedule the best time to repair (Figure 3.11).

In addition, the software helps to determine a *route* for data collection. The positions from where data should be collected can be configured to form an efficient sequence. This sequence or route is then downloaded to the data collector and can then help the operator in the field to determine which measurement position should be taken next. This ensures that all the necessary data are collected in the least possible time and in the same sequence every time.

Besides the route data, off-route vibration measurements made by the data collectors/analyzers for diagnosing machine condition can also be downloaded in the program for history, trending and analysis purposes.

Advantages

They aid in data collection, management and analysis of machinery data.

They can save long-term machinery data that help to compare present and past condition-monitoring data.

They assist in vibration analysis.

They provide user-friendly reports.

Disadvantages

The software programs are expensive, with sometimes almost the same cost as the data collector/analyzer hardware.

They must be configured for individual requirements. A lot of information is required as initial input.

Programs from a particular supplier are not compatible with other brands of data collectors/analyzers and their programs. Currently there are efforts to resolve this, but up to date it remains a problem.

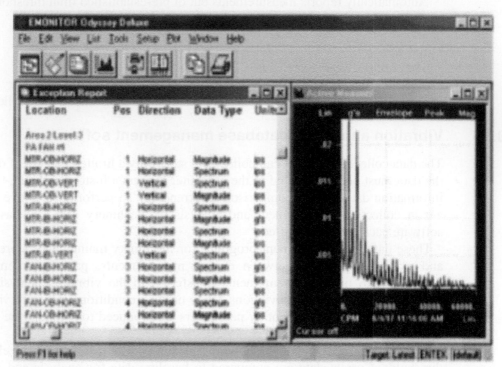

The full range of EMONITOR Odyssey plots provides you with the most complete tools for machinery analysis.

Figure 3.11
Typical database management software output (source: Entek EMONITOR Odessy brochure)

3.3.4   Online data acquisition and analysis

Critical machines, as defined in the earlier topics, are almost always provided with continuous online monitoring systems. Here sensors (e.g. Eddy current probes installed in turbomachinery) are permanently installed on the machines at suitable measurement positions and connected to the online data acquisition and analysis equipment. The vibration data are taken automatically for each position and the analysis can be displayed on local monitoring equipment, or can be transferred to a host computer installed with database management software.

Because monitoring equipment are permanently connected to the sensors, intervals between measurements can be short and can be considered as continuous. This ability provides early detection of faults and supplies protective action on critical machinery. Protective action taken by online data acquisition and analysis equipment is in the form of providing alarms to warn the operators of an abnormal situation (Figure 3.12). In cases of serious faults, this protective action can shut down machines automatically to prevent catastrophic failures.

Transferring the information to a host computer with database management software enhances the convenience and the power of online data acquisition. It is also possible to

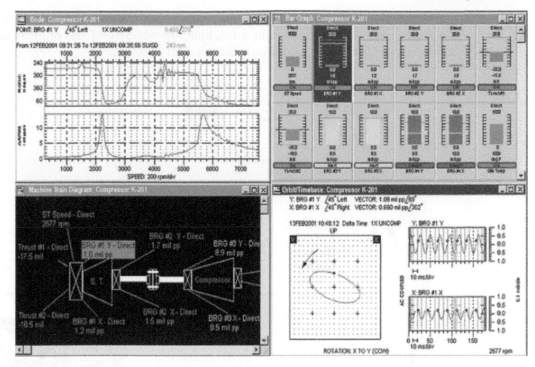

Figure 3.12
Data acquisition software output (source: Bently Nevada – Machinery Asset Management software brochure)

connect multiple local monitoring units that can send data from different machines to a central host computer. Thus, machines at various physical locations can be monitored from one location. Also, information can be transferred from the host computer to the local monitoring unit for remote control (see Figure 3.13).

Vibration analysis/database management software can also be networked to multiple computers with the local area network (LAN) or a wide area network (WAN) to allow multiple users to perform condition monitoring of the machines.

Advantages
Performs continuous, online monitoring of critical machinery.
Measurements are taken automatically without human interference.
Provides almost instantaneous detection of defects.

Disadvantages
Reliability of online systems must be at the same level as the machines they monitor.
Failure can prove to be very expensive.
Installation and analysis require special skills.
These are expensive systems.

3.3.5   Knowledge-based information systems

In addition to the basic software mentioned above, there are also knowledge-based information computer programs available commercially that utilize principles of artificial intelligence (AI). Using the experience of vibration analysts, information from the machine design and its vibration characteristics, these systems automatically analyze and assess the machine condition. After assessing the information, the software can provide a diagnosis of possible problems in a machine, the severity of the problem and can even recommend actions that can alleviate the detected problem.

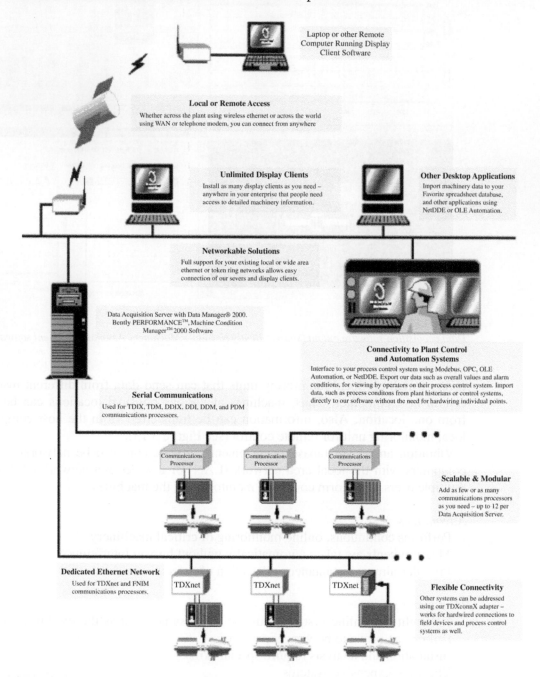

Figure 3.13
A typical online monitoring system (source: Bently Nevada – Machinery Asset Management software brochure)

Often called *expert systems*, these systems require information of the machine being analyzed and its characteristics in order to build a mathematical model of the machine. The model is then stored and used by the expert system to analyze current data and predict a pending problem. The system studies the symptoms of the machine and makes recommendations in order of confidence factors with respect to the severity of the problem.

Expert systems analyze current data and compare it with historical data to search for any changes. The systems then assess the severity of significant changes using absolute thresholds, statistical limits and the rate of change in the calculations. A series of proven rules are then applied to the data. Finally, all rule violations are combined to produce a probability that the diagnosis is correct.

3.3.6 Phase measurement systems

The data acquisition techniques discussed earlier were related to the measurement of vibration motion, and especially related to the amplitude of the vibration. A very important aspect of the vibration wave, next to its amplitude and frequency, is the phase relationships. In vibration analysis, the phase difference between two different waveforms is utilized for many different applications, for example machinery defect diagnosis.

Phase can be measured in many ways, and some are discussed below. It should be noted that in vibration analysis, phase measurement instruments are used in conjunction with the analyzers. The analyzer picks up the vibration waveform, and the following instruments can provide phase and rpm information:

- Stroboscopes
- Dual channel analyzers
- Photocells (phototachometers)
- Electromagnetic and non-contact pickups (keyphasors/shaft encoders).

Stroboscopes

Stroboscopes are normally a part of the accessories kit supplied with a vibration analyzer, but they are available as separate instruments as well. Stroboscopes have a high-intensity light that is flashed at a certain frequency, triggered internally or by the vibration analyzer. It provides a visual method for *observing* phase differences (Figure 3.14).

Figure 3.14
Stroboscope (source: Monarch Stroboscopes)

To obtain a phase difference reading, a reference mark is made on the rotor. A keyway or a notch that can be easily viewed is also a good reference. Another method is to make an angular scale with markings from 0 to 360° on the shaft. This makes phase readings with the smallest error possible. The stroboscope is triggered by the 1× rpm vibration signal from the analyzer, and the reference mark appears to be stationary at some angular position. This angular position is then recorded.

Subsequently, the vibration probe is fixed at another position. The strobe is again triggered by the 1×rpm vibration at this position. The reference mark will now, depending on the situation, appear stationary at the same or some other angular position. This reading is also recorded. The phase difference between the two positions on the machine where the vibration probes were placed is given by the difference of the angular positions of the reference marks as observed with the strobe.

In some cases, only a portion of the shaft may be visible, as just a side view from the coupling guard. In cases where only half of the shaft circumference is visible, a single reference may be inadequate, especially when the reference mark is on the hidden side. In these cases, two marks can be made, for example a small 'o' and 'x' 180° apart.

It should be stated that this method is only reliable for general phase comparisons because it is a visual method and thus approximate. For a more accurate phase reading, a tape marked with angular degrees is applied around the circumference of the shaft. However, even this is a bit cumbersome when the shaft diameter is small.

Advantages

Stroboscopes are lightweight, easy to use and portable.

They can be used individually to measure rpm, by using an internal trigger to make the shaft appear stationary. The frequency of the trigger is the same as the rotational speed of the shaft.

Loose coupling bolts, damaged coupling shims and other defects can be observed by 'freezing' the assembly with strobes.

Some strobes can act as external triggers just like photocells, laser tachometers or keyphasors.

Disadvantages

Machines that do not have a reference notch or keyway must be stopped to provide one. This may be difficult in continuous process plants.

It can only provide phase difference for 1×rpm vibrations. If, for example, the vibration of interest is 2×rpm, the strobe will flash twice during one rotation of the shaft and the reference mark will appear in two positions. This is a problem even with sub-harmonics and higher harmonic vibrations.

It is rather difficult to obtain accurate phase readings in degrees.

Extreme caution is required to use strobes in hazardous areas.

To read the phase, one has to be in close proximity of the machine.

Dual channel analyzers

A single channel analyzer can only accept an input from one accelerometer at a time, whereas a dual channel instrument (Figure 3.15) can accept inputs from two accelerometers simultaneously from different locations on the machine. Thus, two vibration waveforms can be collected from a machine and analyzed. As we shall see later, this can provide very meaningful vibration data.

Advantages

The biggest advantage is that there is no need for reference marks on the shaft. As a result, there is no need to shut down the machine to provide the marks.

The phase differences obtained are very accurate.

It can provide phase differences at any frequency.

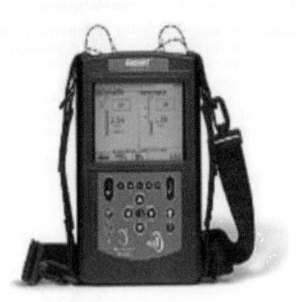

Figure 3.15
Dual channel analyzer

Disadvantages
The dual channel model of analyzer is more expensive than single channel analyzers.

Photocell

Whenever accurate or remote readouts of phase are required, a photocell detector, an electromagnetic or non-contact pickup may be used. These are usually installed within close proximity of the shaft (Figure 3.16). As with stroboscopes, a reference mark on the shaft must be provided to trigger these pickups. A photocell detector responds to the reflectivity of the target. One very common way is to wrap the shaft with a black tape (e.g. electrical insulation tape) and then stick a thin reflective across the tape, or paint a white line on the tape. The objective is to provide an abrupt change in reflectivity of the target area of the photocell during each revolution of the shaft.

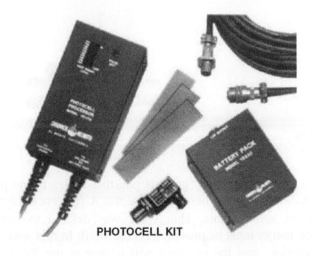

PHOTOCELL KIT

Figure 3.16
Photocell kit

The photocell is somewhat similar in principle to the stroboscope, except that its electronics remove the manual component. A steady light source, rather than a strobo-scopic light source, is transmitted from the device. A photo detector (or photocell) produces a pulse each time light is reflected from a reflective surface on the rotating shaft (the reflective tape). All phase measurements are made relative to the reflective tape, which is treated as zero degrees. Since the reflected light produces one pulse per revolution, the shaft rotation rate can also be determined easily (Figure 3.17).

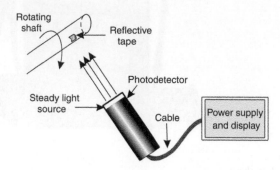

Figure 3.17
Photocell principle

Electromagnetic and non-contact pickups (keyphasor)

With the non-contact probe or electromagnetic pickup, the shaft should have a notch, depression, key or keyway. A temporary method is to attach a key around the shaft with a high-strength tape to hold the key in place. This is *not* recommended for high-speed shafts.

In an electromagnet pickup sensor (Figure 3.18), the output voltage changes to indicate that the reference feature has passed. This output voltage pulse change is then compared with the occurrence of maximum vibration amplitude to determine the phase difference at different locations. The photocell and keyphasor cannot indicate phase readings on their own. They must transmit their data to an analyzer or oscilloscope for analysis.

Figure 3.18
Keyphasor

Unlike with the stroboscope, it is possible to determine phase at 2× rpm and lower and higher harmonics with these instruments. A reference signal at the desired frequency is required to achieve this. Thus, if it was desired to monitor the 2× rpm phase, two reference marks must be provided on the shaft. In this way, two triggers will be generated per revolution, and the analyzer will trigger at the 2× rpm and consequently the phase reading will be at 2× rpm.

Where sub-multiple or non-harmonic-related vibration frequencies must be compared, a reference vibration pickup and a *reference* vibration analyzer with a tuneable filter can be used to provide a reference signal at any desired frequency of vibration.

3.3.7 Torsional vibrations

Torsional vibrations are similar to lateral vibrations that were discussed in the earlier sections. Every mechanical system that rotates and transmits power is subject to some kind of torsional behavior. The three important kinds of torsional behavior that are mostly referred to for analyses are:

1. Changes in revolution
2. Torsional vibration
3. Transmission error.

Torsional vibrometers to measure the above-mentioned phenomena are also described briefly below.

Changes in revolution

For every rotation, any point on the circumference of a shaft has to cover an angular distance of 360° to complete one rotation. The rate at which it rotates is called the angular velocity. Most of us believe that it occurs almost uniformly. However, this is not the case in many situations.

Let us assume the shaft speed is 1 rpm. Let us break the journey of any point on the circumference of the shaft into four sectors, 90° apart from each other. If the angular velocity is uniform, then each sector should be covered in 15 s exactly. Instead, say the first sector is covered in 10 s, the next in 20 s; the third is covered again in 10 s and the final section in 20 s. Thus, in one cycle the angular velocity increases and decreases, giving rise to what is known as changes in revolution.

To measure this change in revolution, a high-resolution rotary encoder is used. The output signal from the encoder is converted from frequency to voltage (F–V) at high speeds, pulse by pulse, to determine the changes in each revolution (Figure 3.19).

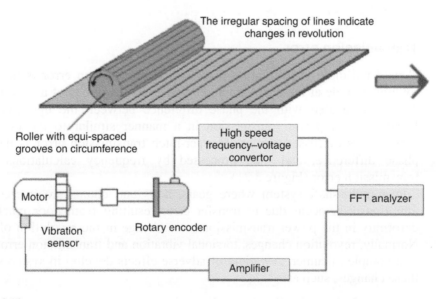

Figure 3.19
Angular velocity vs revolutions

Torsional vibrations

Torsional vibration represents changes in the relative angular displacement between two points on a rotating shaft. Select two collinear points on the circumference of a stationary shaft. When the shaft begins to transmit power, it can twist and then there is no longer a straight line joining the two points. This line could then be slanted or spiral in shape (Figure 3.20).

Normally, a pair of electromagnetic revolution sensors is used for this measurement. The vibrometer measures the torsion that develops between the driving and load sides of the shaft as torsional phase difference. The phase difference obtained is F–V (frequency to voltage) converted at high speeds and processed using frequency calculations to determine the torsional vibration.

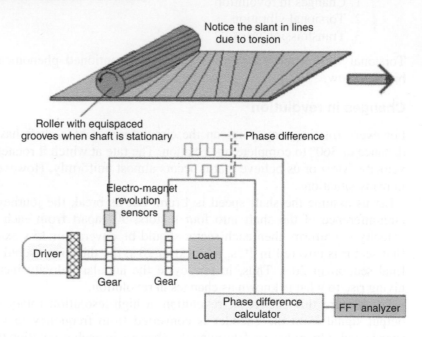

Figure 3.20
Torsional vibration

Transmission error

Consider a multi-shaft-rotating machine. A transmission error is the lead or lag in the rotational angle at the upstream and downstream locations of a power transmission unit. This is measured with the phase difference between the upstream and downstream locations of a power transmission in a manner similar to the detection of torsional vibration as discussed earlier. The per-pitch transmission error is determined from the phase difference, and then processed by frequency calculations to determine the transmission error (Figure 3.21).

In a multi-shaft system where gears, belts and chains are interconnected, torsional vibrations can occur due to transfer error resulting from poor machining accuracy, a deformity in the power transmissions or a change in the revolution of the rotating parts. Normally, revolution changes, torsional vibration and transmission errors are intermingled in a complex manner. A variety of adverse effects develop in systems which experience these changes, such as:

- Increase in vibration and/or noise
- Deterioration in the positioning accuracy

- Deterioration in the feed accuracy
- Breakdowns due to fatigue.

Thus, in any rotating system comprising of rotating parts and power transmissions, it is important to measure these three parameters: revolution change, torsional vibration and transmission error in order to analyze the cause-and-effect relationships.

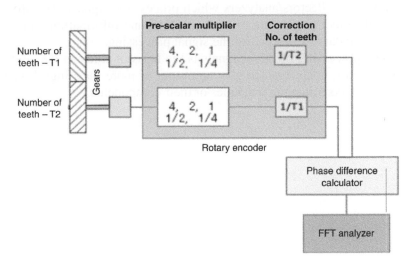

Figure 3.21
Determining transmission error

Torsional vibrometer

Using a dual-sensor system, torsional vibrometers (Figure 3.22) can optically detect rotational pulses, measure changes in the angular displacement and analyze torsional vibrations by simply attaching a striped tape onto the rotating shaft under consideration. The FFT analyzer built into the main unit enables frequency analysis tracking analysis and printing of analysis results in the field.

Figure 3.22
Torsional vibrometer (source: Ono Sokki, Japan Technical Report, website)

In the torsional vibration schematics shown before, assume that it is not possible to measure the relative displacement between two points. The component of torsional vibration can be analyzed, however, from the changes in angular displacement by simply sticking a striped tape at a particular position on the object.

3.4   Conclusion

Thus, this topic encompasses the way in which a machine's vibration motion is captured using transducers and converted into electrical signals. These signals are then collected with data collectors/analyzers which process the signals. Transformation of the electrical signal to its components by the analyzers, and information and documentation related to vibration data form a part of the signal processing.

The processing of these signals is the basis of study in the next topic. After processing the signals, analyzers can give an output of more meaningful data that can be correlated with defects arising in machines or their components.

4

Signal processing, applications and representations

4.1   The fast Fourier transform (FFT) analysis

The vibration of a machine is a physical motion. Vibration transducers convert this motion into an electrical signal. The electrical signal is then passed on to data collectors or analyzers. The analyzers then process this signal to give the FFTs and other parameters. We will take a brief look at the processing of the signals, which finally provide us with the necessary information for condition monitoring. To achieve the final relevant output, the signal is processed with the following steps:

- Analog signal input
- Anti-alias filter
- A/D converter
- Overlap
- Windows
- FFT
- Averaging
- Display/storage.

Before we can discuss the above-mentioned digital signal processing steps, we need to take note of a few more terms and concepts.

4.1.1   Fourier transform

A vibration or a system response can be represented by displacement, velocity and acceleration amplitudes in both time and frequency domains (Figure 4.1).

Time domain consists of amplitude that varies with time. This is commonly referred to as *filter-out* or *overall reading*.

Frequency domain is the domain where amplitudes are shown as series of sine and cosine waves. These waves have a magnitude and a phase, which vary with frequency.

The measured vibrations are always in analog form (time domain), and need to be transformed to the frequency domain. This is the purpose of the fast Fourier transform (FFT). The FFT is thus a calculation on a sampled signal. If FFT is a calculation on a sampled signal, the first question that arises is: how do we determine the sampling rate?

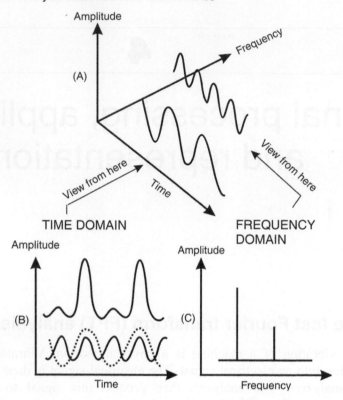

Figure 4.1
Fourier transform

Sampling rate

Sampling is the process of recording the amplitude of a wave at given instants, and then generating a curve from the recorded points. Thus, the collected discrete sampled data points (digital) are used to reconstruct the wave, which was originally in an analog form. If the reconstructed digital wave has to look similar to the original wave, how fast should we record the amplitude, or in other words, take samples so that the digitized wave is an exact replica of the original analog wave?

The answer lies in the Nyquist sampling theorem, which states: 'If we are not to lose any information contained in a sampled signal, we must sample at a frequency rate of at least twice the highest frequency component of interest.'

Figure 4.2 shows a case where the sampling rate is less than double the wave frequency. We can see that four sample intervals collected in 3 ms will result in a reconstructed wave (dotted) as shown in the figure. This wave is of a lower frequency and not at all a true representation of the actual wave.

This phenomenon of formation of a lower-frequency wave due to undersampling is called *aliasing*. All data collectors/analyzers have automatically selected built-in sampling rates to ensure that no aliasing occurs. In theory, there should be no vibrations with frequencies of more than half of this sampling rate. However, this can never be ensured in practice.

Therefore all analyzers are fitted with *anti-aliasing filters*. These are low-pass electronic filters, which allow low frequencies to pass but block higher ones. The filters remove all vibrations in the analog signal that have frequencies greater than half the sampling rate. These filters are automatically tuned to the proper values as the sampling frequency is changed (this occurs when the frequency range of the analyzer is changed by the user). It is very important to note that filtering has to occur before digitisation of the analog commences.

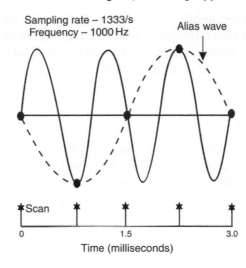

Figure 4.2
Example of undersampling

4.1.2   Analog to digital converters

The vibration waves collected by transducers are analog signals. Analog signals must be converted to digital values for further processing. This conversion from an analog signal to a digital signal is done by an *Analog to Digital (A/D) converter*. The A/D conversion is essentially done by microprocessors. Like any digital processor, A/D conversion works in the powers of two (called binary numbers). A 12-bit A/D converter provides 4096 intervals whereas a 16-bit A/D converter would provide 65 536 discrete intervals (Figure 4.3).

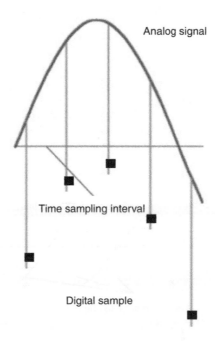

Figure 4.3
Analog to digital converters

The greater the number of intervals, the better is the amplitude resolution of the signal. A 12-bit A/D converter would result in a resolution of 0.025% of the full scale, whereas a 16-bit A/D converter would yield a resolution of 0.0015%. It is thus possible to collect a signal with large and small amplitudes accurately.

4.1.3   Windowing

After the signal was digitized using an A/D converter, the next step in the process (before it can be subjected to the FFT algorithm) is called *windowing*. A 'window' must be applied to the data to minimize signal 'leakage' effects. Windowing is the equivalent of multiplying the signal sample by a window function of the same length.

When an analog signal is captured, it is sampled with fixed time intervals. Sampling fixed time intervals can cause the actual waveform to get truncated at its start and end. The results obtained can vary with the location of the sample with respect to the waveform's period. This results in discontinuities in the continuous waveform. Windowing fills the discontinuities in the data by forcing the sampled data to zero at the beginning and at the end of the sampling period.

Figure 4.4 shows the effects of windowing. Windows can be thought of as a way to fill in the discontinuities in the data by forcing the sampled data to zero at the beginning and end of the sampling period (or time window), thereby making the sampled period appear to be continuous. When the signal is not windowed and is discontinuous, a 'leakage error' occurs when the FFT algorithm is applied.

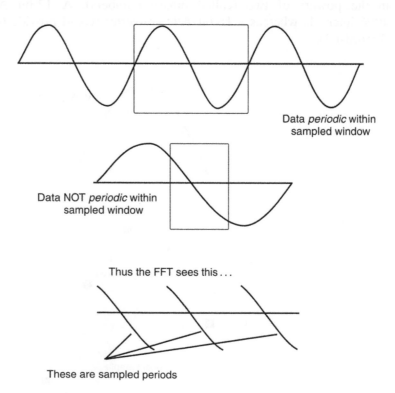

Data *periodic* within sampled window

Data NOT *periodic* within sampled window

Thus the FFT sees this . . .

These are sampled periods

Figure 4.4
The principle of windowing

The FFT algorithm sees the discontinuities as modulating (varying) frequencies and it shows as sidebands in the spectrum when none of these frequencies are actually present in the signal. The usage of windows also affects the ability to resolve closely spaced frequencies while attempting to maintain amplitude accuracy. However, it is possible to optimize one at the expense of the other.

There are many window functions. Some used in vibration signal processing are:

1. Rectangular (basically no window)
2. Flat top
3. Hanning
4. Hamming (Figure 4.5)
5. Kaiser Bessel
6. Blackman
7. Barlett.

Generally, only the first three window functions mentioned above are available in most analyzers.

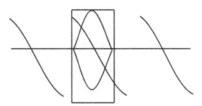

Here a WINDOW function is applied –

example – HANNING WINDOW

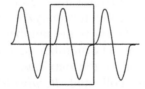

Now the FFT algorithm sees this . . .

Figure 4.5
Window functions

When application of the flat top window is compared to when no window (rectangular window or uniform) is applied, a 'broader' peak is observed in the FFT. The Hanning window also results in *broadening* of the peak, but to a lesser extent than the flat top. Discerning between two very close frequencies becomes very difficult due to the *broadness* of the peaks. When the intent is to identify the presence of a signal component (a peak) at a specific frequency, it is best to apply a rectangular window to do the analysis. But, if the magnitude of the peak is important, the flat top window is clearly the best (Figure 4.6).

Obviously, neither rectangular nor flat top is the best solution. The real solution depends on the purpose of the analysis. For most applications, the best solution actually means processing the data in a number of different ways. For a first result, a Hanning window is optimal and is usually selected by default, as it provides good amplitude resolution of the peaks between bins as well as minimal broadening of the peak.

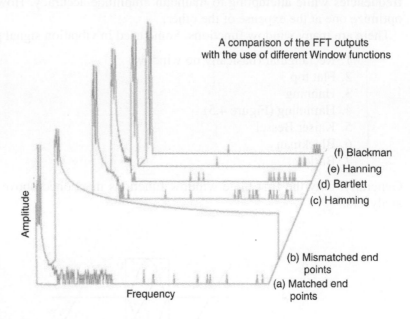

A comparison of the FFT outputs with the use of different Window functions

(f) Blackman
(e) Hanning
(d) Bartlett
(c) Hamming

(b) Mismatched end points
(a) Matched end points

Amplitude

Frequency

Figure 4.6
Comparison of FFT output

4.1.4   Lines of resolution, F-max, bandwidth

After calculation of the FFT on the digital signal, the frequency domain of the signal can be displayed on the collector/analyzer screen. The FFT is a spectrum of amplitude vs frequency. The *resolution* is the number of lines (or bins) that are used to display the frequency spectrum. The number of lines could be 200, 400, 800, 1600, 3200, 6400 and 12 800. *F-max* is the maximum frequency selected on the analyzer by the user when the data are collected. *Bandwidth* is calculated by dividing the F-max by resolution.

It can now be deduced that when resolution is high there is a better distinction between frequency peaks. Selection of F-max upon collecting data requires experience. If F-max is set too high, the bandwidth gets larger and resolution is affected. On the other hand, if the F-max is set too low, valuable high-frequency vibration data could be lost.

Furthermore, some may find it amusing to know that the time required for collecting the data varies inversely with F-max. The higher F-max, the quicker the FFT can be displayed. This is due to a fixed mathematical relationship between sampling rate and the number of bins in the FFT. As a general guideline, the following advice is provided to select F-max values:

- For general rotating machinery like pumps, fans, blowers and motors, set the F-max to 20× or 40×, where × is the running speed.
- When measuring vibrations on gearboxes, the F-max setting should be at least three times higher than the gear mesh frequency, where the gear mesh frequency is the number of teeth of pinion and gear times their respective running speeds.

- However, if an analysis on a machine is conducted for the first time, it is advisable to begin by taking two spectra, one at 10× the running speed and another at 100× running speed. This is to ensure that no important frequencies are lost in the high- or low-frequency zone. Once the range of the suspicious frequencies is noted, the F-max setting should be selected accordingly.

4.1.5   Averaging

Averaging is another feature provided in analyzers/data collectors. The purpose is to obtain more repeatable results, and it also makes interpretation of complex and noisy signals significantly easier. There are various types of averaging:

- Linear averaging
- Peak hold
- Exponential
- Synchronous time averaging.

Linear averaging

Each FFT spectrum collected during a measurement is added to one another and then divided by the number of additions. This helps in obtaining repeatable data and tends to average out random noise. This is the most commonly used averaging technique. The spectra are typically averaged 2, 4, 8, 16 or 32 times, but any number could be used.

Peak hold

With this method, the peak value in each analysis cell is registered and then displayed. In other words, it develops an envelope of the highest spectral line amplitude measured for any average. This technique is used for viewing transients, such as coastdowns or random excitations that may be required during stress analysis studies.

Exponential

In this method, the most recent spectra taken are considered to be more important than older ones, and thus given more mathematical weight when adding and averaging them. This is used for observing conditions that change very slowly with respect to sampling time.

Synchronous time

This method uses a synchronising signal from the machine under investigation, and is used for averaging in the time domain. The synchronising signal is usually in the form of a pulse generated by a photocell or an electromagnetic pickup at a reference position on the shaft circumference. The vibration samples can in this way be taken at the same instant with respect to shaft rotation during averaging.

   Non-synchronous vibrations in the system are effectively nullified by this method. The method is generally used if a machine has many rotational components rotating at different speeds. Thus, the vibrations synchronous with the synchronising signal are emphasized while others are averaged out.

4.1.6   Overlap

Consider the following example: If there is a need to collect and analyze a frequency range of 1 kHz, the data collection time (also known as the time window) for collecting

1024 samples could be exactly 40 ms. The FFT processor (Figure 4.7) can calculate and display a spectrum in 10 ms, after which it encounters an idle duration of 30 ms until the acquisition of the next block is completed.

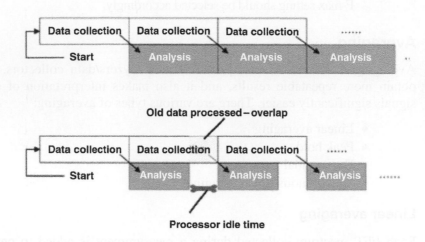

Figure 4.7
FFT data processor

Once the first block is collected, rather than waiting for the next block to be fully collected, it is possible to proceed and calculate a new spectrum by using part of the data from the new block and part of the data from the old block. If the process under consideration is stationary (not varying with time), the data from the two blocks can be averaged.

Considering the example mentioned above, we could initiate a new FFT calculation by using 75% of the previous block and 25% of the new one. We would then be performing a 75% overlap processing and our apparent processing time (after the first block) would be 10 ms per spectrum, rather than 40 ms. The method of overlapping becomes even more significant when we are operating at very low frequencies, or when we want to calculate many spectral averages.

For example, let us assume we are collecting data in a 100-Hz frequency range and wish to calculate 16 averages. The data collection time is 4 s, and without overlap processing we will need 64 s. With 75% overlap, we need 4 s for the first block and 1 s for each successive one, or $4 \times 1 + 1 \times 15 = 19$ s to perform the same task. A considerable amount of time can be saved during data collection by the use of overlapping. The method enables more efficient use of the collected data.

4.1.7   Display/storage

FFT analyzers have LCD screens and built-in memory. This enables it to display the processed signals almost immediately after digitisation. The user can subsequently download this data to the host computer. Some special display modes that are available with modern analyzers are:

- Frequency bands/alarms
- Waterfalls.

Frequency bands/alarms

In this type of display, the FFT spectrum is sub-divided into six regions called bands, each of which has its own alarm limit. This provides the user with the ability to gather information and set alarm limits based on component defects. For example, a region or band of an FFT spectrum can be set up to monitor bearing defect frequencies. There is a trend among some of the vibration tolerances towards filtered vibration limits based on frequencies and harmonics. As a general rule, vibration levels tend to decrease as frequency increases with rotating machinery.

Normally two types of band alarms are used. One is based on the peak amplitude within the band and the other is based on the total destructive energy content within the band. Some of the vibration standards organisations have developed very good alarm guidelines for the different bands of FFT spectra, depending on the type of machinery.

These alarm bands are usually automatically incorporated when loading the machine data in the software associated with the analyzers. This data gets loaded onto the analyzer, along with the route data in which the machine information is supposed to be collected.

Waterfalls

The waterfall is a special display of the FFTs collected on the same position on a machine over a period of time. Each FFT is plotted one after the other giving an impression of a waterfall of FFTs. This kind of display makes it very easy to view variations in amplitudes of any frequency over the entire range. Spectral data stored in analyzers would tend to use too much of the precious memory and it is therefore better to download FFTs to a host computer and use the computer software to display waterfall plots. Examples of waterfall plots are shown in Figure 4.8.

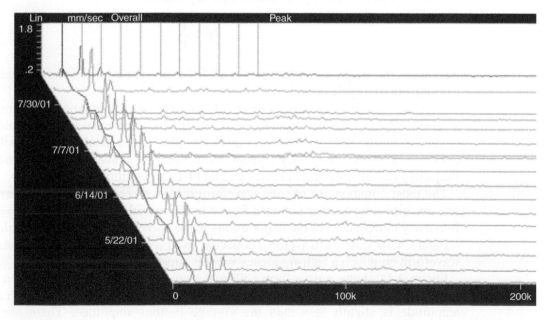

Figure 4.8
Waterfall plot of high-speed steam turbine drive end bearing

The first one is of a blower belt driven by a motor. It has an overhung cantilever type of rotor. The waterfall plots enable a comparison of the present FFT with earlier ones. In this case, the old data were collected nearly two years before the new data. From this plot, it can be easily seen when a rotor was removed for balancing!

The second plot is data from a high-speed steam turbine with water-cooled supports. Peaks developing due to misalignment are clearly visible. Through the waterfall plot, we can keep track of this defect and be warned if the concerned frequencies grow abnormally. As shown, harmonics can be marked on the plot for quick identification.

4.2   Time waveform analysis

A time waveform is the time domain signal. In vibration terms, it is a graph of displacement, velocity or acceleration with respect to time. The time span of such a signal is normally in the millisecond range (Figure 4.9).

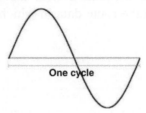

One cycle

Figure 4.9
Time waveforms

Time waveforms display a short time interval of the raw vibration. Though not as versatile as the FFT, it still has particular applications and can provide clues of a machine's condition that may not be evident in the frequency spectrum. The analysis of the time waveform is not a new technique. In the early days of vibration analysis, time waveforms were viewed with oscilloscopes and frequency components calculated manually. The relationship between frequency and time is:

$$f = \frac{1}{T}$$

where f is the frequency in Hz (cpm) and T the time period in seconds, which is required to complete one cycle of the wave.

Thus, reading this raw data on an oscilloscope or on modern analyzers, the cycles over a period of time T can be counted to estimate the frequency that will be obtained. In most of the cases, the time waveform is quite complex and it is almost impossible to count cycles to determine frequency content. However, time waveform study is not meant for frequency calculations and in most cases an FFT calculation is more suitable.

The most common use of time waveform data is to compare the waveform pattern of one machine with another obtained from a machine with similar defects. If necessary, the frequency components of the major events in the waveform pattern can be calculated.

Figure 4.10 shows a waveform collected from a pump with a predominantly 1× rpm waveform on which a high-frequency wave is superimposed.

Figure 4.11 shows a waveform collected from the NDE of double suction ID fan of a boiler. It had harmonics of the fundamental frequency. Note that average positive amplitude is slightly less than the average bottom amplitude. There are also seven countable peaks in the bottom half (or seven cycles) measured in duration of 400 ms. This equals 1050 cycles per minute, and corresponds to the speed of the fan.

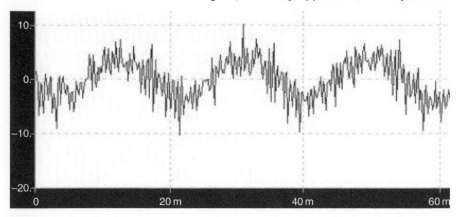

Figure 4.10
Time waveform example

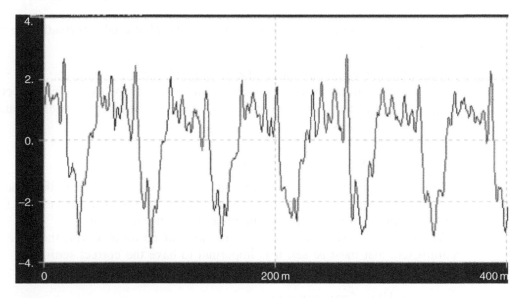

Figure 4.11
Time waveform example

Figure 4.12 shows a special waveform describing a phenomenon called *beats*. Two waveforms having frequencies separated only slightly (less than 30 Hz or so) and with approximately the same amplitude will produce a beating waveform. These are literally pulses due to alternating reinforcement and cancellation of amplitudes. The amplitude change is called the modulation and has a frequency equal to the difference between the frequencies of the two waveforms. If the difference decreases, the beat frequency will also decrease.

Beating waveforms are common at centrifuges that have bowl or scroll at marginally different speeds, and it is very normal to obtain the beat frequency if there is some unbalance in each. In some cases, it might be possible to time the beats to determine the difference between the bowl and scroll speeds.

This phenomenon also occurs in motors that have electrical defects. These defects tend to generate a vibration frequency of twice the transmission power line frequency. If the line frequency is 50 Hz (3000 cpm), the defect frequency would be 6000 cpm. Now, if the motor's physical speed were 2980 rpm, then its second harmonic would be 5960 cpm. Thus, the two waveforms of 6000 and 5960 cpm will generate beats and modulation of amplitude.

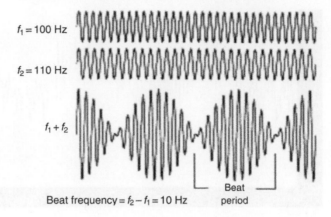

$f_1 = 100$ Hz

$f_2 = 110$ Hz

$f_1 + f_2$

Beat frequency $= f_2 - f_1 = 10$ Hz   Beat period

Figure 4.12
Waveform – beats

Areas where the time waveform can provide additional information to that obtained from FFTs are:

- Low-speed applications (less than 100 rpm)
- Indication of true amplitude in situations where impacts occur, such as assessment of the severity of defects in rolling element bearings and gears
- Looseness
- Rubs
- Beats.

In the case of defects such as unbalance or misalignment, where the time waveform is not too complex, it will not be an advantage to the time waveform for diagnosis. Investigating the FFT amplitude and phase is a better technique.

When a time waveform analysis is carried out on an analyzer, there are a number of parameters that must be set. It is important to have the correct settings to obtain a good *snapshot* of the amplitude of the vibration. These parameters are:

- Unit of measurement
- Time period sampled
- Resolution
- Averaging
- Windows.

4.2.1   Unit of measurement

This refers to the characteristic of vibration of interest, namely displacement in microns (settings are possible for high and low frequencies, defined as μH and μL), velocity in mm/s-pk or acceleration in g. The selection criteria for the units of measurement were discussed earlier in topic 2 under – *Displacement, Velocity, and Acceleration – Which should be used?* It is generally recommended that the natural unit of the transducer be used to display the time waveform.

Thus, if a displacement plot is required, a displacement transducer should be used. In most instances with current commercial data collectors, this would mean that acceleration would be the unit of choice. If data were gathered from non-contact probes on sleeve-bearing machines, displacement is usually used.

4.2.2   Time period of sample

To obtain a usable time waveform for analysis work, the instrument should be set to measure 5–10 cycles of the vibration wave or rotations of the machine being measured. This information is also provided in the *Help* function of some analyzers, which would provide something similar to the table below. The total sample period desired can be calculated by this formula:

Total sample period (ms) = 60 000 × number of revolutions desired/rpm

Machine rpm	Time Period for 5 Revolutions (ms)	Time Period for 10 Revolutions (ms)
2900	103	206
1480	203	406
985	305	608
3600	83	167
1800	167	333
1200	250	500

4.2.3   Resolution

For time waveform analysis, it is recommended that 1600 lines (4096 samples) be used. This ensures that the collected data have sufficient accuracy and key events are captured. Other sampling rates that are normally available are 256, 512, 1024, 2048, 4096 and 8192.

4.2.4   Averaging

In most data collectors, averaging is done during the FFT calculation process. If the synchronous time averaging function is not invoked, the time waveform presented on the screen will be the last one measured even if multiple averages were selected in the analyzer setup. It is therefore normal to take a single average when analysing time waveforms. Overlap averaging should be disabled.

Synchronous time averaging can be used to *synchronize* data acquisition with respect to a particular rotating component. This can be useful when monitoring gearboxes, for instance to locate a defective tooth relative to a reference mark. It is also useful on reciprocating equipment to 'time' events with reference to a particular crank angle.

4.2.5   Windows

As explained in a previous section, various windowing functions can be used to minimize leakage errors while performing the FFT. Some instruments can apply these windows to the time waveform data as well. This would force the data to zero at the start and end of the time sample, and thus some information is lost in the waveform. To eliminate this effect, a uniform or rectangular window should be used.

4.3   Phase signal analysis

Phase signal analysis provides further insight into machine diagnostics in conjunction with the information initially provided by the FFT. For example, the FFT spectrums of an unbalanced and a bent shaft may seem alike. To resolve the problem further, phase analysis is employed. Phase analysis is further used for orbital analysis, rotor balancing, measuring speed and can also aid in the analysis of variable speed machines. Analyzers employ phase analysis to aid diagnostics with the following methods:

- Order analysis/tracking mode
- Orbits
- Resonance identification
- Modal analysis (MA)
- Operational deflection shape (ODS) analysis.

In order analysis, both phase and amplitude are obtained. Order analysis requires an external trigger, which synchronizes both the zero time sample in the time domain buffer and the sample clock such that the 1× component is always the first order. Tracking filter phase errors are canceled out by firmware corrections of the analyzer. Although most phase analyses are performed at 1× rpm, it is *not* restricted to 1× rpm. Phase at 2×, 3× or any other frequency can also be measured. It should hence be emphasized that phase analysis is nothing more than looking at the relative movement of different parts of the machine at a given frequency, and is not restricted to a specific frequency.

The purpose of order analysis is to lock the display to the rotational speed of the machine under test, so that if running speed varies, the ordered components remain in the same position of the display. Thus, the 1× component will not shift when the rpm changes. A phase vs orders display is given, which provides phase relationships with respect to the trigger.

Order tracking is primarily used for carrying out vibration analysis of variable speed machines. The method provides data in the orders domain instead of the time domain. Thus, the *X*-axis of the graph of overall vibrations is not amplitude vs time but amplitude vs orders of rpm (e.g. 10 orders = 10× = 10× rpm). In the orders spectrum, signals that are periodic with shaft revolution appear as peaks and harmonics of shaft rpm remain fixed even if the shaft speed changes. Consequently, spectrum analysis for variable speeds is only possible with the order tracking method. However, this is only possible with the help of an external trigger like a photocell or keyphasor.

In the tracking mode, the trigger not only synchronizes the zero time domain signal but also adjusts the sample clock (as per Nyquist's theorem, the sampling rate has to be at least twice the maximum frequency) to a typical factor of 2.56 of the maximum orders that were selected by the user. Some analyzers allow tracking up to 20× rpm. This ensures that the phase measurements are automatically bin-centered and the accuracy is further enhanced by compensating for the phase error contributions of the tracking filter. A typical order tracking analyzer display would look like:

Tracking Filter	
Test Point:	
Units	Velocity – peak
Speed	2900 cpm
Overall	3.9 mm/s
1 ×	2 ×
Magnitude = 1.2 mm/s	Magnitude = 0.6 mm/s
Phase = 39°	Phase = 81°

Applications for the tracking mode often relate to balancing in order to calculate the influence coefficients (machine-specific information that helps avoid trial runs during balancing). The balancing program calculates machine coefficients by applying known trial weights to balancing plane(s). The system's response to a trial weight is obtained by the vector difference between the data from the reference run and the data from the trial run. Once the program calculates the influence coefficients, it is possible to estimate correction weights or trim weights without further trial runs.

Another application is to ascertain misalignment in machines. Two accelerometers are placed in axial directions on either side of the coupling. The tracking mode can determine if there is a 180° phase shift that is occurring due to the misalignment defect. The other applications such as orbits, resonance identification, modal analysis and operational deflection shapes are discussed in detail later.

4.4   Special signal processes

4.4.1   Synchronous time averaging

Synchronous time averaging is an analysis technique used in locating sources that may be responsible for exciting vibration. This technique collects time waveforms synchronized to a 1× 'marker' pulse. These synchronous time waveforms are then averaged in the time domain, and the resultant frequency spectrum of this averaged waveform is displayed. Synchronous time averaging is used to measure vibrations directly *and* harmonically related to the turning speed of a specific component or shaft.

Synchronous time averaging is often used when there are more than one shaft on a machine that are turning at different speeds, or if there are several machines in close proximity to each other. It is possible to use synchronous time averaging to remove the vibrations of non-synchronous components from the signal, so that only the vibration from the reference shaft is recorded.

Synchronous averaging requires a tachometer or any other device to provide a 1× pulse. These pulses are used to start data acquisition in synchronisation with the reference shaft rotation. The shaft on which the tachometer is located is the *reference* shaft. All the machine vibrations are recorded during each data acquisition *window*. The frequencies not synchronously related to the speed of the reference shaft are effectively removed from the signal through averaging.

Rotor-related vibration problems such as imbalance, misalignment, rotor looseness and rubs are retained in the spectrum display provided by synchronous time averaging. Defects not synchronous with the rotating shaft such as bearing faults, cavitation, electrical noise and resonance produce non-synchronous frequencies that are effectively averaged out of the spectrum. In this way, synchronous time averaging is an effective diagnostic tool for isolating specific faults.

Synchronous time averaging displays peaks synchronously related to the reference shaft. Since synchronous time averaging is an averaging technique, a number of averages may be required to remove non-synchronous peaks. It is important to know that synchronous time averaging takes many averages to allow non-synchronous peaks to average out of the spectrum. At times, 100–1000 averages are needed to identify vibration components related to the reference shaft.

Synchronous time averaging can be performed using an analyzer to display the synchronous peaks related to the reference rotor. As an example, consider the belt-driven fan illustrated in Figure 4.13. As shown in the illustration, the machine consists of two parallel belts connected to shafts turning at different speeds. Both the motor and the fan shafts are closely tied to the same support frame.

- Belt frequency   685 cpm
- Motor, 1× rpm   1485 cpm
- Fan shaft, 1× rpm   1000 cpm
- Motor, 4× rpm or electrical?   5964 cpm.

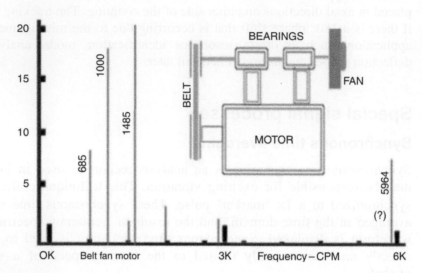

Figure 4.13
Normal FFT spectrum

Synchronous time averaging was used to identify the vibration originating from each shaft. Figure 4.14 shows the synchronous time averaged signature using a trigger on the motor shaft. The vibration components originating from the motor shaft are all that remain after synchronous time averaging.

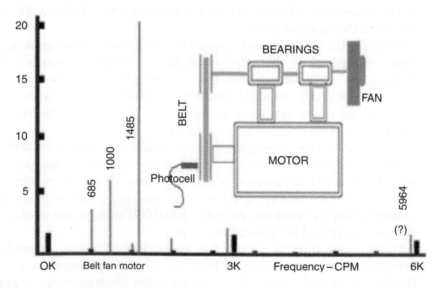

Figure 4.14
Synchronized time-averaged spectra with trigger on motor shaft – note the drop in amplitude of other frequencies not synchronous with the motor

Figure 4.15 is the spectrum originating from synchronous time averaging on the fan shaft. Again, all vibration components have been averaged out of the spectrum except those originating from the fan shaft. It is interesting to note that two specific frequencies averaged out of both synchronous time-averaged spectrums. The belt frequency, 685 cpm was averaged out because it was not synchronous with either shaft. The 5964-cpm peak also averaged out of both spectra.

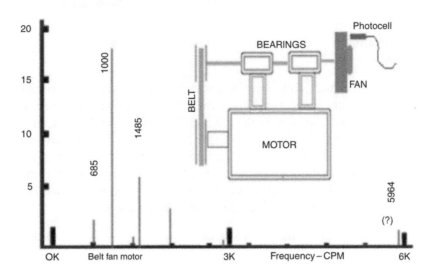

Figure 4.15
Synchronous time-averaged spectra with trigger on fan shaft

The 5964-cpm peak could have been confused with 4× the motor speed. Had this been the case, the peak would not have averaged out of the synchronous time-averaging spectrum. Since 5964-cpm peak did average out, it can be concluded that the cause is non-synchronous and could be an electrical problem in the motor stator winding.

4.4.2 Orbits

Orbits are Lissajous patterns of time domain signals that are simultaneously plotted in the X–Y coordinate plane of an oscilloscope or vibration analyzer. In this form of display, it is very difficult to trace the start of the orbit as it appears to be an endless loop. In order for us to determine the direction of rotation, a phase trigger is employed. The trigger will show the direction of rotation by looking at the dot on the orbit as the starting point of 1× rpm and the blank space as the end point.

Orbit analysis is the vibration measure of any rotor system in an X–Y plot (Figure 4.16). In most applications, the unit of measurement is displacement which is measured directly using proximity probes. These types of measurements are relative vibration readings. Relative readings are considered vibration measurements of the shaft with respect to the bearing housing. As the probes are clamped firmly to the housing, there is no relative motion between the probe and the housing. Thus, the orbit is achieved. With that in mind, orbit plots give a visual graph of the actual shaft centerline movement inside the bearing housing.

Accelerometers and velocity pickups can also be used to create orbits. These are external transducers, which require mounting on the outside of the bearing housing. These types of measurements are called case orbits. Case orbits are useful to separate shaft and case vibrations. This can provide absolute shaft motion (relative to space).

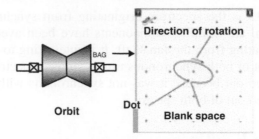

Figure 4.16
Orbit analysis

To understand orbits, waveforms and their relationship to orbits must be explained. Let us begin with waveforms. The waveform plot shown in Figure 4.17 has two sine waves, Y and X. The Y plot is on the top and the X plot is at the bottom. The waveform signature runs left to right and the amplitudes change from negative to positive, whatever the case may be. The changes in the waveform cause the orbit to form. An orbit is made up of an X- and Y-axis with zero in the center. Starting from the center, up is positive and down is negative. Right is positive and left is negative. Now that we know waveform and orbit conventions, let us trace the waveforms and create an orbit.

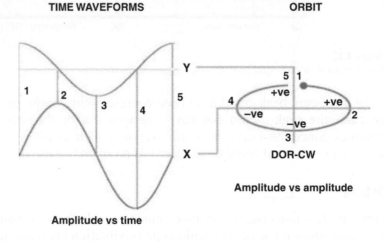

Figure 4.17
Waveforms and its relationship to orbits

If the Y waveform were to be traced, we can see it starts at point 1 where its amplitude is a maximum of say a unit of 25 microns, and goes to zero at point 2. It reaches −25 microns at point 3, zero again at point 4 and finally +25 microns at point 5. This cycle continues up and down. If this trace existed alone, it would describe a straight line from +25 microns to −25 microns.

The X waveform starts with zero amplitude at point 1, which happens to be 90° behind Y. Here, the X waveform travels through the 5 points with maximum amplitude of 50 microns, for instance. The X waveform alone would produce a horizontal line from +50 microns to −50 microns. Now consider both the waveforms as they move through time. This will combine the X and Y point in the orbit to complete the ellipse.

To visualize the orbit, look at the plot and notice that the Y is at maximum amplitude when X is zero. At point 1, the Y waveform is at maximum amplitude and the X waveform is zero. As we move to point 2, which is 90° to the right, the orbit X is at maximum amplitude at +50 microns when Y is zero. At point 3, again 90° to the right, the X waveform is zero and the Y waveform is at −25 microns. At point 4, a further 90°, the X waveform reaches −50 microns but now the Y waveform is zero. Point 5 is the same as point 1.

The orbit will change as the amplitude and phase angles change in the waveforms. As mentioned earlier, the dot on the orbit signifies the start of the wave and the blank space denotes the end. In this example of an orbit, as we moved from left to right, points 1–5 on the orbit move in a counter-clockwise direction, following the convention. If the shaft rotation also happens to be moving in a clockwise direction as it traversed from the Y probe to the X probe, the orbit would be considered to be in *forward precession*, otherwise it would be in *reverse precession.*

Bode plot

A Bode plot comprises of two graphs:

- Amplitude vs machine speed
- Phase vs machine speed.

To display a Bode plot, a phase trigger is used to obtain a shaft reference for phase measurement and measure the machine speed. The analyzer triggers and records the amplitude and phase simultaneously at specific speed intervals (which can be defined by the user), and the two graphs are displayed on top of each other.

In rotordynamics, the Bode plot is mainly used to determine the critical speed of the rotor. In the plot, the speed at which amplitude of vibration is maximum is noted, and for confirmation the phase graph is checked to see if it differs from the starting value by 90° (Figure 4.18).

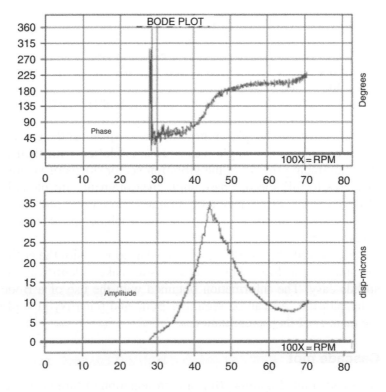

Figure 4.18
Bode plot showing 'critical speed' of rotor at approximately 4400 rpm. Note that at this speed, amplitude is a maximum at 35 microns. The phase, which at 2900 rpm was 45°, is 135° at 4400 rpm indicating a difference of 90°

The Bode plot can also be used to determine the amount of runout associated with a proximity probe, the balance condition, system damping and the operational phase angle cum amplitude at various machine speeds.

Polar/Nyquist plot

The polar or Nyquist plot is also a representation of the same three variables as considered in a Bode plot. The variables are plotted on a single circular chart instead of Cartesian axes (Figure 4.19).

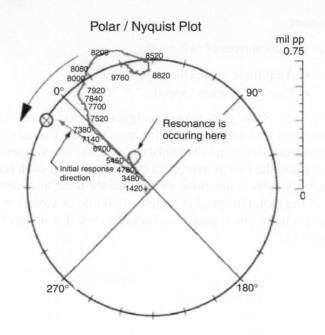

Figure 4.19
A Nyquist plot indicating critical speed of a rotor (resonance) at approximately 5000 rpm

The center of the plot represents zero speed and zero amplitude with subsequent amplitude and phase angle measurements plotted with their associated machine speed. It is like tracing the tip of the vibration vector (locus of vector tip) as the machine speed increases to full speed. Phase angle measurements are plotted around the circumference of the chart, against machine rotation direction. Polar plots are always filtered to machine running speed or some multiple of the machine speed, depending upon the fault being investigated.

Machine critical speeds are displayed as loops, with the critical speed situated 90° from the start of the loop. This characteristic makes identification of resonance and critical speeds easy. The information obtained with the use of a Bode plot is possible with the polar plot too. As can be seen, the Bode and polar/Nyquist plots are the same except in the manner of their presentation.

Cascade plot

A cascade plot (Figure 4.20) is a representation of three parameters: amplitude, frequency and machine speeds. An FFT plot of amplitude vs frequency is recorded at specific machine speed intervals (selected by the user). After collection of all the FFTs, they are cascaded one after another in a form similar to a waterfall plot.

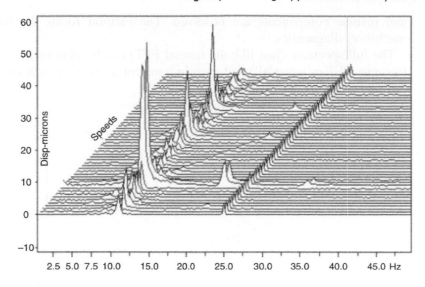

Figure 4.20
Cascade plot

It is important to note that a waterfall plot is the FFT of the same location collected at different time intervals. Waterfall plots are of a machine running at the same speed. A cascade plot, however, is a collection of FFTs at different machine speeds and is taken during transition of a machine speed, for instance during a start-up or a coastdown (shutdown). In the cascade plot, orders (1/2×, 1× or higher) are displayed in the form of a line.

It is evident that a cascade plot is a tool for transient analysis, which forms an essential diagnosis tool for critical machinery. The Bode and Nyquist plots also belong to this group of transient analysis tools.

Full spectrum

The full spectrum is an additional diagnostic tool and is also called the spectrum of an orbit. It shows the same information as an orbit but in a different format. It helps to determine the degree of ellipticity (or flattening) associated with the various machinery conditions along with the precessional direction for all the frequency components present.

To obtain the full spectrum, the orthogonal X and Y transducer signals are fed into the direct and quadrature parts of the FFT input. The positive and negative vibration components for each frequency are obtained. Positive is defined to be the forward precession and the negative component as the reverse precession. These components yield the following ellipticity and precessional information for a given orbit of any particular frequency (1× or 2× or …):

- The sum of two components, forward and reverse, is the length of the orbit major axis.
- The difference between the two components is the length of the orbit minor axis.
- The larger of the two components, positive or negative, determines the direction of precession that is forward or reverse.

One of the possible applications of full spectrum is analysis of the rotor runout caused by mechanical, electrical or magnetic irregularities. Depending on the periodicity of such irregularities observed by the X–Y proximity probes, different combinations of forward and reverse components are observed. The method forms the basis for *many* useful machinery diagnostics.

The full spectrum (just like the normal FFT) can be obtained in a steady-state analysis (a single FFT or waterfall) and even in transient analysis, which would then be called the full spectrum cascade (Figure 4.21).

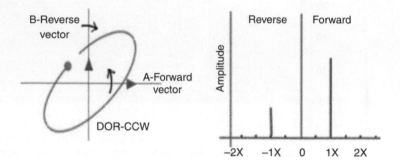

Figure 4.21
A steady-state single full spectrum

4.4.3   Operational deflection shapes analysis

Operational deflection shape (ODS) analysis (Figure 4.22) is a technique for visualisation of vibratory movement of a machine under its operating conditions. This method is really helpful to track down the root cause of defects, or show the machine behavior to persons who are not familiar with vibration analysis.

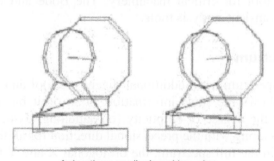

Animations as displaced by using
operating deflection shapes (ODS) analysis

Figure 4.22
ODS analysis

In this method, a picture of the vibratory motion of the machine is constructed by measuring the vibration and phase data of various locations on the machine. The relative amplitude and phase information of the various positions is measured and these data are given to a host computer that has special software to simulate the response data. The end result is an actual, but highly exaggerated picture of the motion of the different locations on the machine.

ODS analysis is the term used for this visualisation technique. It relies on the machine's actual vibratory movement while operating to provide information. For this reason, it is relatively a straightforward technique to apply and does not rely on approximate models of loads and structures. The data should be collected with the machine *operating as it is*.

ODS analysis is possible with a single channel analyzer with a synchronous once per revolution reference pulse. However, ODS is best performed by using a dual channel analyzer. An advantage is that no shutdowns of machines are required to perform an ODS analysis.

ODS visualisation software packages for animating the machine movement are commercially available, and are usually dedicated to the task. However, an existing standard modal analysis software package can also be used. This does not imply that a modal analysis must be carried out, only that the ODS is substituted for a computed *mode shape* for the purpose of animation. Software is particularly useful when there are many monitoring positions on the machine.

ODS vs modal analysis

It must be emphasized that ODS differs from modal analysis (MA). The major differences are listed below:

* Modal analysis is commonly used for evaluating structural designs and investigating design modifications before structural changes are made.

  The ODS is used to analyze and solve complex operational machinery problems, by providing an animation of its motion.
* Modal analysis requires specially controlled and measured loads to be applied to the machine structure, preferably while it is not operating, and then uses the vibration measurements to calculate the structure's dynamic behavior. Using this information, MA creates a mathematical model of mass, stiffness and damping characteristics of the structure and determines its mode shapes at its natural frequencies. These properties are independent of the operating conditions of the structure.

  The ODS, on the other hand, is quite simple and relies totally on the actual operating loads of the machine to excite vibration. The vibration measurements then provide a visualisation of the vibratory movement under operating conditions. With different loading or operating conditions, different deflection shapes will occur.

Measurements for ODS are collected at specific frequencies (1×, 2× or any other frequency) using a vibration spectrum analysis to determine the amplitude at the frequencies at which the ODS is to be constructed. One measurement location is identified as a reference position and the relative vibration levels and phases at the chosen frequency are measured at all other locations using a *roving* sensor.

The relative amplitude and phase values are sufficient, and absolute values are of no consequence. If required, absolute values could always be indexed with a fixed position. Including the total effect of multiple frequencies, synchronous or asynchronous, or perhaps all frequencies below a certain cut-off, can accommodate more complex vibration shapes, although the latter is less common. Vibration normally occurs in all directions and ODS analysis can be conducted for 2 or 3 directions of the movement if desired.

ODS with single channel analyzers

As stated above, provided there is a synchronous reference pulse available to trigger the analyzer's data acquisition, many single channel analyzers and some data collectors can measure the magnitude and phase of synchronous vibrations (Figure 4.23).

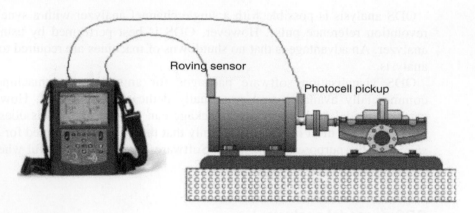

Figure 4.23
ODS with single channel analyzers

Relative magnitude is calculated by dividing the measured magnitude at each location by the magnitude of the vibration at the reference position. The phase measured in this case is the phase with respect to the trigger. Relative phase is the difference between the phase with respect to the trigger at a given location and the phase with respect to trigger at the reference location.

ODS with multi-channel analyzers

There is no need for triggered measurements when using a multi-channel analyzer, because the relative amplitude and phase can be measured directly and accurately using the transfer function between the two channels (Figure 4.24).

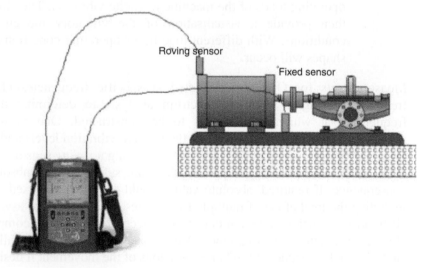

Figure 4.24
ODS with multi-channel analysers

A simple approach is to make cross-channel measurements between the vibration at the reference location and the vibration at every other location in turn. The frequency component to be used may be selected from a cross-spectrum magnitude plot.

The relative magnitude and phase follow from the transfer function plot of the reference location to the current location.

Some of the new terminology mentioned here will be explained in the following topics.

Cross-spectrum

A cross-spectrum is basically a comparison between the signals from two points of interest on a machine. It essentially consists of two graphs of:

- Amplitude vs frequency
- Phase vs frequency.

The amplitude vs frequency plot shows the relationship between the magnitudes of the two signals. For instance, if at 1450 cpm both signals have visible peaks, it will also be seen on the cross-spectrum at 1450 cpm. If only one signal had a 1450-cpm peak, it will not show in the cross-spectrum.

The phase vs frequency is a plot of the phase difference between the two signals as a function of frequency. The terminology used for a cross-channel spectrum with two input channels labeled A and B is either:

- The cross-spectrum from A to B
- The cross-spectrum from B to A.

This depends on whether we want to relate the phase of channel B to channel A, or the phase of channel A to channel B, respectively.

The magnitudes in each case will be the same, but the phase difference will have opposite signs. The magnitude of the cross-spectrum is the product of the magnitudes of the two instantaneous spectra. The phase is the difference between the phase of channel B and channel A. Thus, if two signals have identical amplitude and phase at a particular frequency, the cross-spectrum magnitude will be equal to the *power spectrum* of the signal (mean square amplitude spectrum, not the rms). The phase at that particular frequency in the cross-spectrum will be zero.

Frequency response function (FRF)

The frequency response function (FRF) for a system describes the relationship between input and output of the system as a function of frequency, where the input is usually force and the output acceleration. The FRF gives the magnitude of the output per unit of input and relative phase between output and input, as a function of frequency. The FRF describes a cause-and-effect relationship.

Transfer response function (TRF)

The transfer response function (TRF) compares two outputs of a system related to a common input. The outputs are usually acceleration. The TRF gives the magnitude of the two outputs and the relative phase between them. There is not necessarily a cause–effect relationship between the outputs.

Coherence

The coherence function, or coherence (or sometimes called coherency) is a measure of the degree of linear relationship between two signals as a function of frequency. In other words, coherence is an indication of the linearity of a system at any frequency. It has a similar role in frequency domain analysis to the correlation coefficient in statistical analysis.

Coherence has magnitude values of zero to one, where zero means no coherency and one means total coherency. For a single channel, coherence is one at all frequencies, because of the way the function is defined. To be useful, therefore, coherence measurements must be combined with cross-channel analysis preferably with averaging turned on.

Coherence measurement is thus only possible with a dual channel analyzer and it is a very standard feature on most analyzers.

Consider this example. A transducer is mounted on a blower having vibrations suspected to be originating from its foundation. Another transducer is mounted on the foundation where the vibration is at its highest. Thus, one sensor is measuring the suspected 'cause' and the other is measuring the 'effect'. Coherence can show that, at a particular frequency, the two signals are interdependent or independent. The coherence plot looks almost like a spectrum (Figure 4.25). The frequencies that are interrelated will have a value of nearly one, while frequencies that are unrelated will be much lower, often near zero.

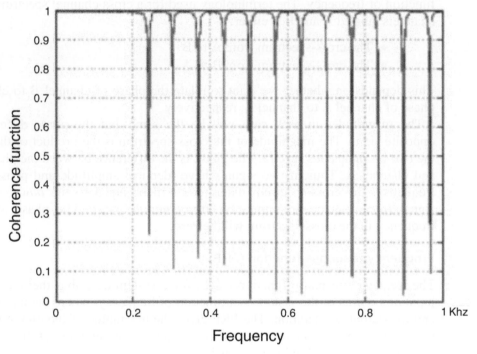

Figure 4.25
A typical coherence output

The mathematics of coherence

It is important to understand the difference between the two types of spectra (Figure 4.26):

1. *Power spectrum:* It is the square of the magnitude of the averaged spectrum of a signal. The phase information is not considered.
2. *Cross-spectrum:* It is the averaged vector product of the spectra of two signals (considering both the phase and magnitude effects).

Coherence is designated by the symbol γ^2.

Coherence between two signals A and B is calculated with the following formula:

$$\gamma^2 = \frac{\left(\text{magnitude of cross-spectrum from A to B}\right)^2}{\left(\text{power spectrum of A}\right)^2 \times \left(\text{power spectrum of B}\right)^2}$$

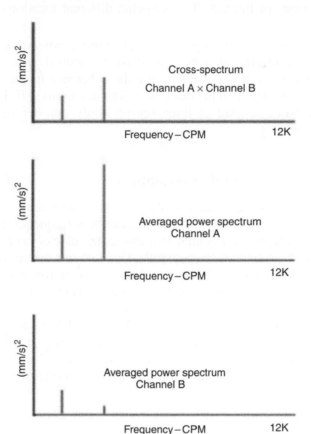

Figure 4.26
The three power spectra required calculating coherence

Applications of coherence

A few practical applications of coherence are discussed below.

Application 1 One of the ways coherence can be used is to identify the source of a vibration that seems unrelated to anything in the machine under consideration. One sensor is fixed at the 'effect' and another roaming sensor is moved around (with a long cable), attaching it to possible 'sources'. The coherence should be checked at each location. The location that yields the maximum coherence at the problematic frequency is in all probability the source of the unwanted vibration.

The same approach can be used when installing a piece of sensitive machinery to ensure that there is not a noticeable transmission of vibration from a nearby machine or transportation system.

Application 2 Another interesting application of coherence is to select the location of sensors for an optimum transmission path, either when trying to identify a suitable place for a permanent mounting position or when seeking the best location for a routine periodic measurement. In these cases, one sensor is placed at a possible mounting position. The machine or structure can be dynamically excited (e.g. with impact hammer) as close as possible to the loading position, source of vibration or vibration energy input.

The input load and the vibration at the receiving sensor can then be measured and coherence determined. A suitable location will have a close-to-unity coherence at most frequencies of interest. By comparing different locations, the optimum selection can be made.

Application 3   A third possible application is when the number of sensors required for online monitoring on a machine must be reduced. Instead of two or three orthogonally placed sensors, it is possible to use the coherence function to determine the transmission of vibrations through the machine casing in that area. If the coherence is high, it might be possible that instead of three sensors, only one sensor could be used in an oblique direction.

4.4.4   Enveloping and demodulation

This technique of vibration analysis is extensively used for fault detection in bearings and gearboxes. This method focuses on the high-frequency zone of the spectrum. Using a high-pass filter (allows high frequencies but blocks lower ones), the analyzer zooms into the *low-level high-frequency data*. The analyzer essentially tries to pick up some peaks that would otherwise be lost in the noise floor (or at times called the carpet, which is nothing but extremely small amplitudes across the whole frequency range of the spectrum) of a narrow band spectrum.

To understand the concept of enveloping and demodulation, let us consider the example of a defective bearing which has a single spalling (a piece of metal chipped off) on the track of the outer race (Figure 4.27). Every time a ball passes through this spall, it would generate an impact, let us call it a 'click' for simplicity.

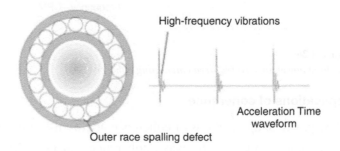

Figure 4.27
Single spalling

For example, let us say in this bearing 6.3 balls pass over this spall defect during every revolution. Then 6.3 clicks per revolution are generated. If the shaft were rotating at 1000 rpm we would see a peak of 6300 cpm on the FFT. This peak may be visible in a regular FFT, but generally the low-frequency area in spectrums is crowded and it may be difficult to notice the small peak and relate it to the bearing or gearbox defect. For this reason, the enveloping technique is used.

Returning to the ball bearing travelling over the spall, we should note that along with 6300-cpm low-frequency peak, there is something else that is happening. Every 'click' of the ball bearing passing over the defect is analogous to someone ringing a bell with a gong. When someone rings the bell, we hear two kinds of sounds. One is of the gong frequency and the other is the ringing sound of the bell. The ringing sound is the resonance of the bell, which is basically a high-frequency vibration. Let us carry this analogy over to the defective ball bearing.

Every 'click' of the ball passing over the defect is similar to the gong hitting the bell. Just as the bell rings, the bearing resonates in a similar manner, generating high-frequency vibrations. The ringing high frequency of the bearing is dependent on the mass and the stiffness of the bearing.

Thus, the frequency of 6300 cpm in the FFT is the same as the gong-hitting frequency, and the resonance of bearing components relates to the ringing of the bell. To understand this technique, we need to look at the high-frequency ringing vibrations of the bearing.

As the balls pass over the spall defect, the bearing continues to generate the ringing frequency. At the instant of the ball impact on the defect, the vibration amplitude reaches maximum value, after which it begins to die out until the next ball impacts and the amplitude rises again. This process continues endlessly, giving rise to a waveform that looks like the one depicted in Figure 4.28.

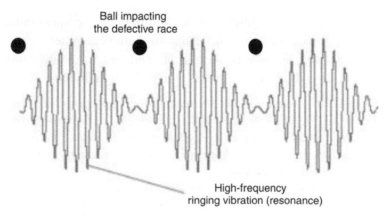

Figure 4.28
Vibration – balls pass over the spall defect

Amplitude modulation due to balls passing over a defective race

This phenomenon of amplitude rising and falling with time is called amplitude modulation in wave terminology. The next step is *demodulation*. In this process, the ringing of the bearing is removed and the process leaves us with a trace that looks like the wave shown in Figure 4.29.

Figure 4.29
Envelope

The trace left after the high-frequency portion of the wave was removed

If this enveloped wave were analyzed, it would appear on the FFT plot as an amplitude peak with the frequency of 6300 cpm, which happens to be the 'click' frequency! Thus, by looking at a small bandwidth in the high-frequency region, we can trace a low-frequency wave. It is important that this low-frequency vibration be impacting in nature to cause amplitude modulation of a high-frequency wave.

This is exactly the kind of vibration generated from defective bearings and gears. Thus, the method of enveloping/demodulation is used for analysing bearings and gears.

Commercially, there are mainly four techniques of enveloping/demodulation, available from CSI, Entek, SKF and SPM instruments. These are used for analysing defects in the high-frequency range. These techniques are known as PeakVue, Spike Energy (gSE), Spectral Emission Energy (SEE) and the Shock Pulse Method (SPM).

Each of these is briefly discussed below.

PeakVue (trademark – Computational Systems Incorporated, US)

The PeakVue method offered by CSI can be briefly described as:

- A high-pass filter removes low-frequency content in the acceleration signal.
- It passes through the analog-to-digital converter.
- Each sample is scrutinized and only the amplitude level that exceeds a specified trigger level is assigned a digital value.

That is, if the analyser is set up to take 1024 samples from the analogue signal, there will be 1024 digital peak values if there are strong impacts. The digital time waveform will only have the highest positive values, one per sample. If there are no defects, there will be no high pulses, and only low amplitude noise (from the signal or instrument) may show.

A pulse from an impact due to a defect bearing or gear has very short time duration and can be periodic; thus there will only be high peak values in the cells or samples where the pulse occurred.

- The FFT algorithm then processes this digital information and the resulting spectrum will only show a fundamental peak and harmonics that have a frequency equal to that of the pulse.

With PeakVue, the signal is extracted from high-frequency data, but there is no rectifier or enveloping done by a low-pass filter.

The gSE spectrum (trademark – Entek IRD International Corporation)

Spike Energy, also called the gSE spectrum, is done as follows:

- A high-pass filter removes the low-frequency content (long time period data) in the acceleration signal.
- It passes through a rectifier, which has an RC time decay to emphasize the impact events.

The filtered signal (step 1) passes through a peak-to-peak detector, which not only holds the peak-to-peak amplitude but also applies a carefully selected decay time constant. The decay time constant is directly related to the spectrum maximum frequency (F-max). The output signal from a Spike Energy peak-to-peak detector is a saw-tooth-shaped signal.

- Even though the rectified signal has only positive values, it is made up of many periods of high-frequency values. It is then 'enveloped' or smoothened electronically and passed through the analog-to-digital converter.
- After the analog-to-digital converter digitized the acceleration signal, the digital values represent the total energy in each time sample, not just the peaks. If there were no defects, there will be no outstanding pulses, perhaps only signal/instrument noise of low amplitude (energy) in each sample.
- To remove the high-frequency content, the digitized information is sent through a low-pass filter. The resulting waveform is the shell or the envelope of the modulation.
- The data is then subjected to a peak-to-peak detector that determines how often the pulse is occurring (the fundamental period), and also determines the duration of the pulse. Thus, when digital information is processed using the FFT algorithm, the resulting spectrum will only show a fundamental peak and harmonics that have the pulse frequency.

The differences between the two are:

- Spike Energy uses some form of rectifier.
- In Spike Energy, the signal is enveloped.
- The amplitudes are not true peak acceleration values. They are gSE (Spike Energy) values.

Spectral emission energy – SEE (trademark – SKF Condition Monitoring)

A spectral emission energy (SEE) device takes high-frequency acoustic measurements and combines them with enveloping techniques to diagnose bearing condition. During enveloping, the vibration signal is filtered to leave only the high frequencies, which removes most of the vibrations caused by structural vibration, misalignment and other factors. The remaining defect signals are reduced in energy content, but are still occurring at the same time intervals. High-frequency vibrations are generated each time a defect bearing is rolled over.

The SEE provides detection of metal-to-metal contact when the lubrication film is broken due to a bearing defect or other similar faults. This contact generates high frequencies in the region of 250–350 kHz. The SEE method monitors bearings by using high-frequency acoustic emission detection using a wideband piezoelectric accelerometer.

The SEE method differs from normal spectral analysis that only go as high as 20 kHz and of enveloping techniques at 5–60 kHz. In this case, an acoustic emission transducer is used. The transducer's contact with the machine should be free of grease or any air gaps that could cause loss of signal strength.

A signal of 250–350 kHz is filtered and enveloped. A low-pass filter then ensures that only low-frequency components remain. This enveloped signal is analyzed digitally using normal analysis equipment.

Shock Pulse Method – SPM (trademark – SPM Instruments)

Shock pulse method which is a patented technique for using vibration signals measured from rotating rolling bearings as the basis for condition monitoring of machines. This technique was invented in 1969. On the impact of a roller with a race defect, the shock pulse transducer reacts with a large amplitude oscillation to the weak shock pulses, because it is excited at its natural frequency at 32 kHz. Machine vibration, at a much lower frequency, is filtered.

The analysis comprises of the following stages:

- The vibration signal from the machine, with superimposed transients at the resonance frequency caused by shock pulses, is passed to an electronic filter.
- The filter passes a train of transients at 32 kHz. Their amplitudes depend on the energy of the shock pulses.
- The transients are converted into analog electronic pulses.
- The converted shock pulse signal from the bearing is finally converted to a rapid sequence of stronger and weaker electrical pulses.

The filtered transducer signal reflects the pressure variation in the rolling interface of the bearing. When the oil film in the bearing is thick, the shock pulse level is low, without distinctive peaks. The level increases when the oil film is reduced, but there are still no distinctive peaks. Damage causes strong pulses at irregular intervals. Shock pulse meters measure the shock signal on a decibel scale, at two levels.

4.4.5   Cepstrum analysis

The one characteristic of vibration spectra common to anti-friction bearings or gears is that there is a kind of harmonic series with the possibility of multiples of the fundamental bearing tones and/or rotational rate (frequency of defect or harmonic, thus add/subtract 1× of the shaft).

This can be described as a common frequency spacing separating the peaks of signature groups. Cepstrum analysis is the name given to a range of techniques involving functions which can be considered as the 'spectrum of a logarithmic spectrum'. It converts a spectrum back to the time domain, and hence has peaks corresponding to the period of the frequency spacings common in the spectrum. These peaks can be used to find the bearing wear frequencies in the original spectrum.

Spectrums from a rotating machine (Figure 4.30) can be quite complex, containing several sets of harmonics from rotating parts and there may be several sets of sidebands due to frequency modulations (changes in frequency mostly due to torsional oscillations).

Because cepstrum has peaks corresponding to the spacing of the harmonics and sidebands, they can be easily identified.

Significant peaks in the cepstrum correspond to possible fundamental bearing frequencies. Using a set of embedded rules, an expert system can automatically compare these frequencies to the peaks in the spectrum that are not related to any machine fundamental forcing frequency. If a match is found, then the spectral peak is considered to be a possible bearing tone, and it is passed to another part of the expert system for rule-based decision-making.

Through cepstrum analysis, the expert system has the advantage of detecting rolling contact bearing wear without knowing exactly what type of bearings the machine uses. Cepstrum also distinguishes bearing wear patterns from flow noise or cavitation.

The word cepstrum is an anagram for spectrum.

4.4.6   Third octave analysis

Third octave spectral analysis is used in analysing acoustical data (Figure 4.31). With this method, a proportional rather than a constant bandwidth is displayed. In a sense, the spectral values are computed over bandwidths that increase with frequency. This is achieved by using a series of bandpass filters covering the various frequency ranges. The ratio of upper to lower band edge frequency is exactly two, so consecutive filters cover one full octave of the frequency spectrum.

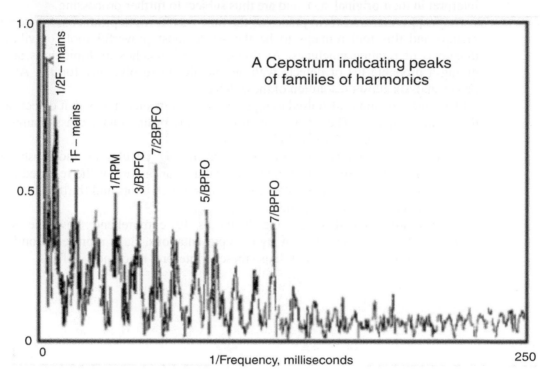

A Cepstrum indicating peaks of families of harmonics

1/Frequency, milliseconds

Figure 4.30
Spectrums from a rotating machine

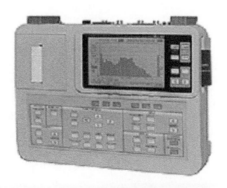

Figure 4.31
Real-time octave band analyzer (source: Ono Sokki, Japan website)

Third octave analysis provides little diagnostic advantages above a conventional spectrum analysis. Third octave data displays are important mainly because many machinery manufacturers and government regulatory agencies specify acceptance criteria in third octaves. Users who are required to test for and comply with OSHA, FAA, ISO or other noise specifications need third octave data display capabilities. *A weighting* of third octave data is used to emulate the response of the human ear.

4.5   Conclusion

There are basically only two types of raw signals that are emitted from machines. These are the time waveforms and the phase reference. These signals are very difficult to interpret in their original state and are thus subject to further processing.

The fast Fourier transform processes the time waveform to produce a frequency domain picture and this tool remains to be the single most powerful tool for vibration fault diagnosis of rotating machines. At times, the FFT reaches its limitations and ends up giving a probable list of defects with an inability to resolve any further. At this stage, phase analysis allows resolution of these defects.

Phase information is also used to represent *signals* in such a way that defects are easy to detect and diagnose. These representations include orbits, ODS, cross-channel analysis, FRF, TRF, coherence, Bode and Nyquist plots.

It was also shown how synchronous time averaging, enveloping and cepstrum analysis assist with 'cleaning' the signal output so that a clear picture of the required information can be formed. Special representations like the cascade plots and third octaves have their own importance for certain applications.

Understanding signal processing helps us to comprehend how the signals are transformed to generate different signal representations. Each representation has its own niche and in the next topic we will use these to interpret machinery defects.

5

Machinery fault diagnosis using vibration analysis

5.1   Introduction

Present day requirements for enhanced reliability of rotating equipment are more critical than ever before, and the demands continue to grow constantly. Advances are constantly made in this area, largely due to the consistent demand from the hydrocarbon, power-generation, process and transportation industries.

Due to the progress made in engineering and materials science, rotating machinery is becoming faster and lightweight. They are also required to run for longer periods of time. All of these factors mean that the detection, location and analysis of faults play a vital role in the quest for highly reliable operations.

Using vibration analysis, the condition of a machine can be constantly monitored. Detailed analyses can be made to determine the health of a machine and identify any faults that may be arising or that already exist.

In this chapter, further attention is given to the method of correlating rotating machine defects to vibrations collected and displayed by the various types of analyzers.

5.2   Commonly witnessed machinery faults diagnosed by vibration analysis

Some of the machinery defects detected using vibration analysis are listed below:

- Unbalance
- Bent shaft
- Eccentricity
- Misalignment
- Looseness
- Belt drive problems
- Gear defects
- Bearing defects
- Electrical faults
- Oil whip/whirl
- Cavitation
- Shaft cracks

- Rotor rubs
- Resonance
- Hydraulic and aerodynamic forces.

We will now look at each one of the above cases in detail and see how they manifest in vibration analysis.

5.2.1 Unbalance

Vibration due to unbalance of a rotor is probably the most common machinery defect. It is luckily also very easy to detect and rectify. The International Standards Organisation (ISO) define unbalance as:

> *That condition, which exists in a rotor when vibratory, force or motion is imparted to its bearings as a result of centrifugal forces.*

It may also be defined as: The *uneven distribution of mass about a rotor's rotating centerline.*

There are two new terminologies used: one is *rotating centerline* and the other is *geometric centerline.*

The *rotating centerline* is defined as the axis about which the rotor would rotate if not constrained by its bearings (also called the principle inertia axis or PIA).

The *geometric centerline* (GCL) is the physical centerline of the rotor.

When the two centerlines are coincident, then the rotor will be in a state of balance. When they are apart, the rotor will be unbalanced. There are three types of unbalance that can be encountered on machines, and these are:

1. Static unbalance (PIA and GCL are parallel)
2. Couple unbalance (PIA and GCL intersect in the center)
3. Dynamic unbalance (PIA and GCL do not touch or coincide).

Static unbalance

For all types of unbalance, the FFT spectrum will show a predominant 1× rpm frequency of vibration. Vibration amplitude at the 1× rpm frequency will vary proportional to the square of the rotational speed. It is always present and normally dominates the vibration spectrum (Figure 5.1).

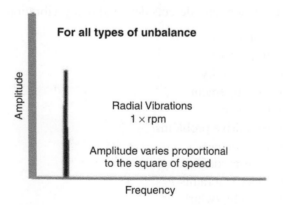

Figure 5.1
FFT analysis – unbalance defect

Static unbalance will be in-phase and steady (15–20°). If the pickup is moved from the vertical (V in the figure) direction to the horizontal (H in the figure) direction, the phase will shift by 90° (±30°). Another test is to move the pickup from one bearing to another in the same plane (vertical or horizontal). The phase will remain the same, if the fault is static unbalance (Figure 5.2).

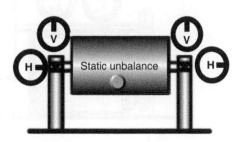

Figure 5.2
Phase relationship – static unbalance

If the machine has no other major defects besides unbalance, the time waveform will be a clean SHM waveform with the frequency the same as the running speed.

Couple unbalance

In a couple unbalance (Figure 5.3) the FFT spectrum again displays a single 1× rpm frequency peak. The amplitude at the 1× varies proportional to the square of speed. This defect may cause high axial and radial vibrations. Couple unbalance tends to be 180° out of phase on the same shaft. Note that almost a 180° phase difference exists between two bearings in the horizontal plane. The same is observed in the vertical plane. It is advisable to perform an operational deflection shape (ODS) analysis to check if couple unbalance is present in a system.

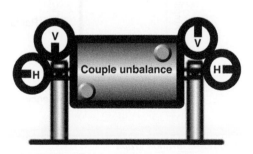

Figure 5.3
Phase relationship – couple unbalance

Unbalance – overhung rotors

In this case, the FFT spectrum displays a single 1× rpm peak as well, and the amplitude again varies proportional to the square of the shaft speed. It may cause high axial and radial vibrations. The axial phase on the two bearings will seem to be in phase whereas the radial phase tends to be unsteady. Overhung rotors can have both static and couple unbalance and must be tested and fixed using analyzers or balancing equipment (Figure 5.4).

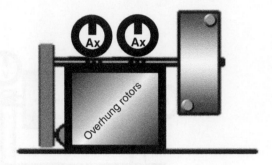

Figure 5.4
A belt-driven fan/blower with an overhung rotor – the phase is measured in the axial direction

5.2.2   Eccentric rotor

Eccentricity occurs when the center of rotation is at an offset from the geometric centerline of a sheave, gear, bearing, motor armature or any other rotor. The maximum amplitude occurs at 1× rpm of the eccentric component in a direction through the centers of the two rotors. Here the amplitude varies with the load even at constant speeds (Figure 5.5).

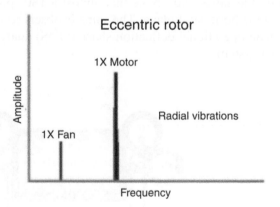

Figure 5.5
A belt-driven fan/blower – vibration graph

In a normal unbalance defect, when the pickup is moved from the vertical to the horizontal direction, a phase shift of 90° will be observed. However in eccentricity, the phase readings differ by 0 or 180° (each indicates straight-line motion) when measured in the horizontal and vertical directions. Attempts to balance an eccentric rotor often result in reducing the vibration in one direction, but increasing it in the other radial direction (depending on the severity of the eccentricity) (Figure 5.6).

Eccentric rotor

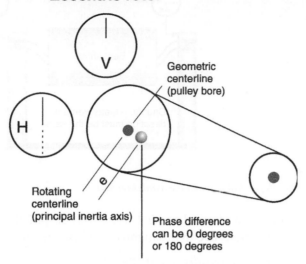

Figure 5.6
Eccentric rotor

5.2.3 Bent shaft

When a bent shaft is encountered, the vibrations in the radial as well as in the axial direction will be high. Axial vibrations may be higher than the radial vibrations. The FFT will normally have 1× and 2× components. If the:

- Amplitude of 1× rpm is dominant then the bend is near the shaft center Figure 5.7)
- Amplitude of 2× rpm is dominant then the bend is near the shaft end.

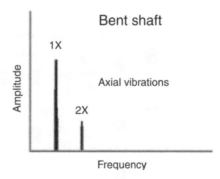

Figure 5.7
An FFT of a bent shaft with bend near the shaft center

The phase will be 180° apart in the axial direction and in the radial direction. This means that when the probe is moved from vertical plane to horizontal plane, there will be no change in the phase reading (Figure 5.8).

5.2.4 Misalignment

Misalignment, just like unbalance, is a major cause of machinery vibration. Some machines have been incorporated with self-aligning bearings and flexible couplings that

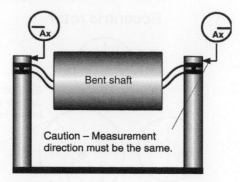

Figure 5.8
Note the 180° phase difference in the axial direction

can take quite a bit of misalignment. However, despite these, it is not uncommon to come across high vibrations due to misalignment. There are basically two types of misalignment:

1. *Angular misalignment:* the shaft centerline of the two shafts meets at angle with each other
2. *Parallel misalignment:* the shaft centerline of the two machines is parallel to each other and have an offset.

Angular misalignment

As shown in Figure 5.9, angular misalignment primarily subjects the driver and driven machine shafts to axial vibrations at the 1× rpm frequency. The figure is an exaggerated and simplistic single-pin representation, but a pure angular misalignment on a machine is rare. Thus, misalignment is rarely seen just as 1× rpm peak. Typically, there will be high axial vibration with both 1× and 2× rpm. However, it is not unusual for 1×, 2× or 3× to dominate. These symptoms may also indicate coupling problems (e.g. looseness) as well (Figure 5.10).

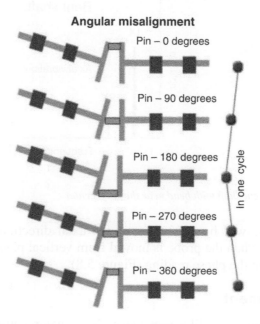

Figure 5.9
Angular misalignment

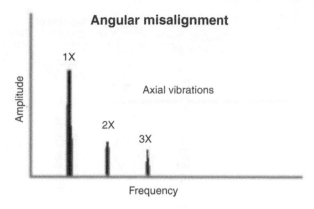

Figure 5.10
FFT of angular misalignment

A 180° phase difference will be observed when measuring the axial phase on the bearings of the two machines across the coupling (Figure 5.11).

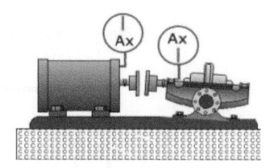

Figure 5.11
Angular misalignment confirmed by phase analysis

Parallel misalignment

Parallel misalignment, as shown in Figure 5.12, results in 2 *hits* per cycle and therefore a 2× rpm vibration in the radial direction. Parallel misalignment has similar vibration symptoms compared to angular misalignment, but shows high radial vibration that approaches a 180° phase difference across the coupling. As stated earlier, pure parallel misalignment is rare and is commonly observed to be in conjunction with angular misalignment. Thus, we will see both the 1× and 2× peaks. When the parallel misalignment is predominant, 2× is often larger than 1×, but its amplitude relative to 1× may often be dictated by the coupling type and its construction.

When either angular or parallel misalignment becomes severe, it can generate high-amplitude peaks at much higher harmonics (3× to 8×) (Figure 5.13) or even a whole series of high-frequency harmonics. Coupling construction will often significantly influence the shape of the spectrum if misalignment is severe (Figure 5.14).

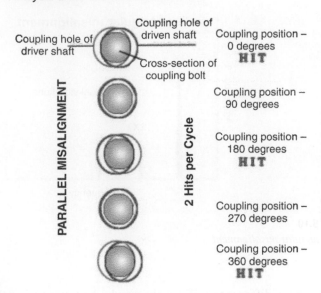

Figure 5.12
Parallel misalignment

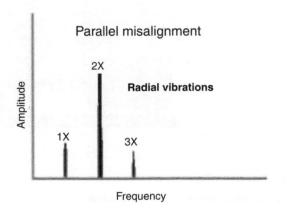

Figure 5.13
FFT of parallel misalignment

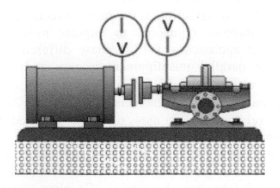

Figure 5.14
Radial phase shift of 180° is observed across the coupling

Misalignment vs bent shaft

Often, a bent shaft and dominant angular misalignment give similar FFT spectrums. The vibrations are visible in both the axial and radial vibration measurements. It is only with phase analysis that these problems can be resolved further. In a machine with a bent shaft, a phase difference will be noticed on the two bearings of the same shaft. In the case of misalignment, the phase difference is visible on bearings across the coupling.

Misaligned bearing cocked on shaft

Misalignment does not only appear with couplings. Often, bearings are not accurately aligned with the shaft. Such cocked bearings can generate considerable axial vibration. A twisting motion is caused with approximately 180° phase shift from the top-to-bottom and/or side-to-side when measured in the *axial direction* of the same bearing housing (Figure 5.15).

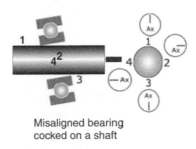

Misaligned bearing
cocked on a shaft

Figure 5.15
Misaligned bearings

Even if the assembly is balanced, high axial vibrations will be measured. The FFT taken in the axial direction will show vibration frequencies of 1×, 2× and 3× rpm.

Attempts to align the coupling or balance the rotor will not alleviate the problem. The cocked bearing must be removed and correctly installed.

In the case of a radial sleeve bearing, no vibrations will be observed due to this cocked assembly. The problem must be accompanied by an unbalance. A radial and axial vibration will be observed, which results from the reaction of the misaligned bearing to the force due to unbalance. Balancing the rotor will reduce vibration levels in both directions.

If a misalignment is suspected, but cannot be confirmed after checking for couplings and cocked bearings, then it becomes necessary to investigate for a condition known as 'soft foot'. This phenomenon will be discussed in a following section.

Viewpoint: There are differences of opinion among experts in connection with the generation of 2× rpm and higher harmonics in the FFT spectrum due to misalignment.

One school of thought agrees with the previous discussion. However, another hypothesis is that the waveform due to misalignment gets modified due to the loads generated by misalignment. Contorted waveforms such as this show 2× and higher harmonics of the fundamental frequency of 1× rpm when subjected to the FFT algorithm.

Misalignment and other radial preloads (orbit representation)

In the previous section, we have seen how the signals from two proximity probes installed on a sleeve bearing (typically on a turbo machine) help to generate an orbit.

An orbit is the trace of the shaft centerline for a single rotation. However, if we were to take an average position of the shaft centerline over a *certain* period of time, the result will describe something like what is shown on the left part of Figure 5.16. If the direction of rotation (DOR) of shaft is clockwise (CW) and if it is normally loaded, the ideal position of the average shaft centerline should be around the 7 o'clock to 8 o'clock position.

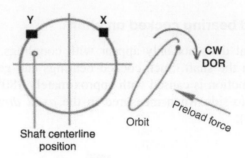

Figure 5.16
Orbit plot due to misalignment

When radial preloads due to misalignment, gravity, fluid forces and other causes increase in magnitude, the orbit will become acutely ellipsoid. A bearing preload due to a cocked assembly can also cause the orbit to have lower amplitude in one axis that makes the ellipse look thinner. The average shaft centerline will move from the normal position to the upper left quadrant, for example.

In the figure, all points on the orbit are moving clockwise (which is the same as the direction of rotation) and therefore the orbit is still in *forward precession*. If the preloading increases further, it will result in the orbit's shape to resemble a number 8 character. In this case, it is also interesting to follow the average shaft centerline position, which has now moved further upwards into the left quadrant. If this orbit is carefully studied, it will be noticed that if a point on the orbit begins its journey from the *dot*, it is moving counter-clockwise initially, whereas the shaft is rotating in the clockwise direction. Thus, heavy preloading due to misalignment can cause the shaft to go into *reverse precession*. Forward precession is normal, reverse is not. If the trajectory of our imaginary point on the trace of the orbit is continued, one can visualize that precessions keep changing continuously (Figure 5.17).

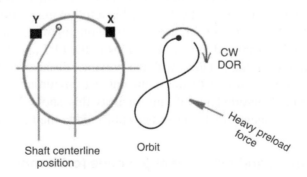

Figure 5.17
Orbit when the preload due to misalignment increase

It should be noted that heavy preloads do not normally cause perfect number 8 orbits, but cause differently sized loops. This kind of loading, typically in a high-speed turbomachine, can be quite damaging and can result in excessive bearing wear, shaft fatigue and possibly shaft cracking.

5.2.5 Mechanical looseness

If we consider any rotating machine, mechanical looseness can occur at three locations:

1. Internal assembly looseness
2. Looseness at machine to base plate interface
3. Structure looseness.

Internal assembly looseness

This category of looseness could be between a bearing liner in its cap, a sleeve or rolling element bearing, or an impeller on a shaft. It is normally caused by an improper fit between component parts, which will produce many harmonics in the FFT due to the non-linear response of the loose parts to the exciting forces from the rotor. A truncation of the time waveform occurs, causing harmonics. The phase is often unstable and can vary broadly from one measurement to the next, particularly if the rotor alters its position on the shaft from one start-up to the next.

Mechanical looseness is often highly directional and may cause noticeably different readings when they are taken at 30° increments in the radial direction all around the bearing housing. Also note that looseness will often cause sub-harmonic multiples at exactly ½× or ⅓× rpm (e.g. ½×, 1½×, 2½× and further) (Figures 5.18 and 5.19).

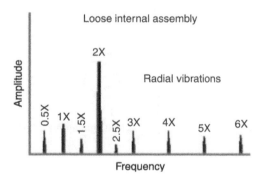

Figure 5.18
Loose internal assembly graph

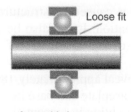

Figure 5.19
Loose fit

Looseness between machine to base plate

This problem is associated with loose pillow-block bolts, cracks in the frame structure or the bearing pedestal. Figures 5.20 and 5.21 make it evident how higher harmonics are generated due to the rocking motion of the pillow block with loose bolts.

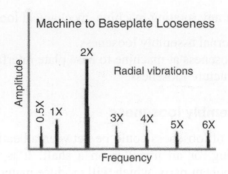

Figure 5.20
Mechanical looseness graph

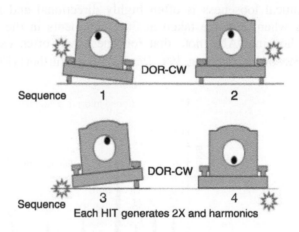

Figure 5.21
Mechanical looseness

Structure looseness

This type of looseness is caused by structural looseness or weaknesses in the machine's feet, baseplate or foundation. It can also be caused by deteriorated grouting, loose hold-down bolts at the base and distortion of the frame or base (known as 'soft foot') (Figure 5.22).

Phase analysis may reveal approximately 180° phase shift between vertical measurements on the machine's foot, baseplate and base itself (Figure 5.23).

When the soft foot condition is suspected, an easy test to confirm for it is to loosen each bolt, one at a time, and see if this brings about significant changes in the vibration. In this case, it might be necessary to re-machine the base or install shims to eliminate the distortion when the mounting bolts are tightened again.

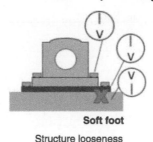

Soft foot

Structure looseness

Figure 5.22
Structure looseness

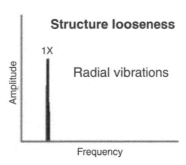

Structure looseness

1X

Radial vibrations

Amplitude

Frequency

Figure 5.23
Structure looseness graph

5.2.6   Resonance

Any object has a natural frequency which is determined by its characteristics of mass, stiffness and damping. If a gong strikes a bell, the bell rings at its own characteristic frequency known as its natural frequency. The gong-striking event is *forced vibration*, whereas the ringing of bell is *free vibration*.

A free vibration at a natural frequency is called resonance.

There is a simple method to find the natural frequency of any object or system called the *bump test*. With this method, a vibration sensor is fixed to the body whose natural frequency is required. Using an impact hammer, a blow is struck on the body and the time waveform or FFT is collected. The dominant frequency observed in the two graphs is the natural frequency of the body. Figures 5.24 and 5.25 show the time waveform and the FFT spectrum of a bump test conducted on a metal study table, respectively.

As seen in the time waveform, the impact occurs at approximately 100 ms after data collection was initiated. Directly after the impact, the body exhibits free vibrations at its own natural frequency. The amplitude of the vibration reduces logarithmically due to damping effects. The period between 500 ms and 1 s is long enough to count the number of cycles. The calculation indicates that the natural frequency is approximately 990 cpm. To obtain the FFT, the data collector was reset and another impact was made on the table with a hammer. The collected spectrum shows a dominant peak at 1046 cpm. This is close to the value calculated before with the time waveform.

The bump test is simple and used extensively in practice. It is a quick and accurate way of finding the resonance frequencies of structures and casings. It is tempting to use the bump test on a spare pump or other rotors not supported on bearings to obtain an estimate of their critical speeds. Take note that this can be very inaccurate. For example, the critical speed of rotors with impellers in a working fluid and supported by their bearings differs vastly from the critical speed obtained using a bump test off-line on the rotor.

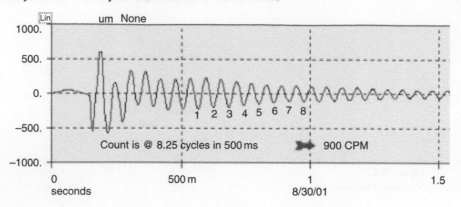

Figure 5.24
Time waveform of a bump test

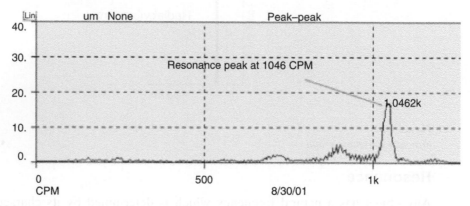

Figure 5.25
FFT spectrum of a bump test

Assume that a multistage pump rotor has a natural frequency of 2500 cpm when pumping a fluid. Assume that the rotor has a slight unbalance, which generates tolerable amplitudes of vibration at 1× rpm. In this example, the unbalance causes the forced vibration frequency at 1× rpm. When the pump is started, the speed begins to increase and along with it also the amplitude and frequency of the vibration due to unbalance. At a particular instant, the forced frequency of vibration due to unbalance will be 2500 cpm. This frequency also happens to be the natural frequency of the rotor. Whenever the forced vibration frequency matches the natural frequency of a system, the amplitude rises significantly, much higher than expected compared to unbalance effects. This condition is called a *critical speed*.

Rotor critical speeds are confirmed using a Bode plot as shown in Figure 5.26. As the rotor approaches its critical speed, the amplitude rises. It reaches a maximum and then drops again. The phase changes steadily as well and the difference is 90° at the critical speed and nearly 180° when it passed through resonance.

The high-vibration amplitudes at critical speeds can be catastrophic for any system and must be avoided at all costs. Besides the example of the natural frequency of a rotor, structural resonance can also originate from support frame foundations, gearboxes or even drive belts.

Natural frequencies of a system cannot be eliminated, but can be shifted to some other frequency by various methods. Another characteristic of natural frequencies is that they remain the same regardless of speed, and this helps to facilitate their detection.

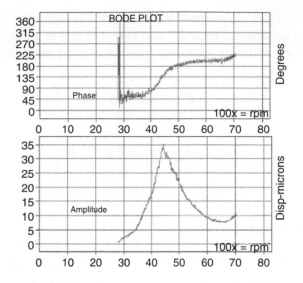

Figure 5.26
Rotor response represented by Bode plot

To comprehend the reason for the shape of the Bode plot, we must go back to vibration basics. We have discussed earlier how the response of an object or mechanical system due to an exciting force such as unbalance can be represented as:

$$Mu \cdot r \cdot \omega^2 \cdot \sin(\omega t) = M(a) + C(v) + k(d)$$

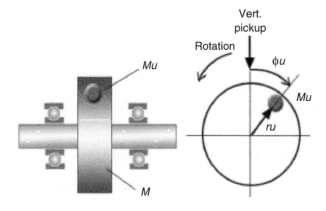

Figure 5.27
A simple rotor system

Consider Figure 5.27, where:

M = rotor mass
Mr = unbalance mass
ru = radius of unbalance
Φu = angular position of unbalance
ω = angular velocity of rotor.

Figure 5.27 shows a simple rotor mass that is supported between two bearings. The mass M has an unbalanced mass Mr at a radius ru and at an angle Φu from the vertical sensor. If the rotor is forced to rotate, the synchronous response will be given by the equation:

$$Mr \cdot a + C \cdot v + K \cdot d = Mu \cdot ru \cdot \omega^2 \cdot \cos(\omega t - \Phi u)$$

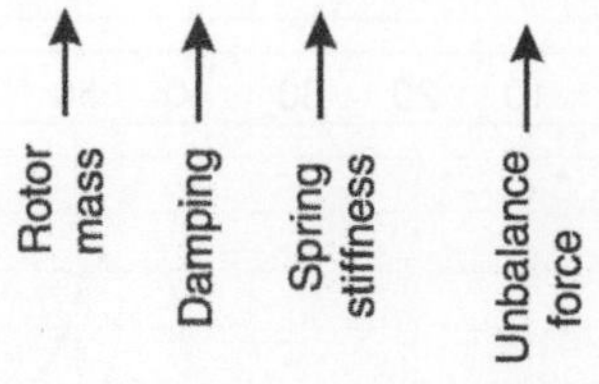

Rotor mass — Damping — Spring stiffness — Unbalance force

The above equation in a vector format would appear graphically as shown in Figure 5.28. In this case, the exciting force is the unbalance force. The mass, damping and stiffness, which are the restraining forces add to a single force called the system force. In order to simplify this equation of motion, we replace acceleration with $d \cdot \omega^2$ and velocity with $d \cdot \omega$.

The previous equation is now modified to:

$$(- Mr \cdot \omega^2 + C \cdot \omega + K) \cdot d = (Mu \cdot ru \cdot \omega^2) \cos(\omega t - \Phi u)$$

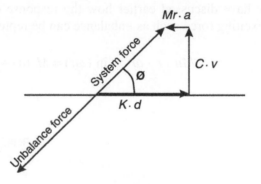

Figure 5.28
System force-graph

Hence, with reference to Figures 5.29 and 5.30:

synchronous dynamic stiffness × displacement = unbalance force

Rearranging the above equation we get:

$$\text{Synchronous response} = \frac{(M \cdot ru \cdot \omega^2)}{(-Mr \cdot \omega^2 + C \cdot \omega + k)}$$

$$\text{Synchronous response} = \frac{\text{unbalance force}}{\text{synchronous dynamic stiffness}}$$

Figure 5.29
SDS Formulas

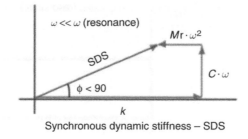

Figure 5.30
SDS graph

Consequently, the synchronous response indicates that the 1× amplitude (displacement) increases when the unbalance force increases (or when the synchronous dynamic stiffness decreases).

We will now investigate the rotor response in the three speed ranges.

Case 1 – Running speed ω is much less than critical speed

When the speed is less than the critical speed, the mass and damping contributions to stiffness are small. The dominant stiffness is the spring stiffness, which keeps the amplitudes low. At these low speeds, the unbalance force is changing and the spring stiffness is assumed not to be changing. The response of the rotor increases quadratic proportional to the speed (Figure 5.31).

The phase relationship of the rotor reference and the heavy spot is such that the vibration lags (phase difference = ϕ) behind the unbalance (heavy spot) and at this stage it is less than 90°.

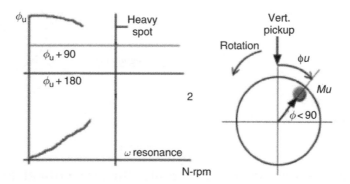

Figure 5.31
Rotor response vs speed increase

Case 2 – Running speed ω is equal to the critical speed (Figure 5.32)

When the rotor running speed approaches the critical speed, the mass stiffness and the spring stiffness contributions to the equation are equal in magnitude but opposite in direction. They thus cancel each other out and the only factor restraining the force is the damping. This is the reason why the synchronous rotor response (displacement at 1×) is at its maximum at the critical speed.

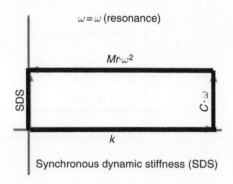

Figure 5.32
Rotor speed reaches critical speed

The phase relationship between the response and the heavy spot is now exactly 90° (Figure 5.33). During critical speed we observe that the vectors:

$$k - M \cdot \omega^2 = 0$$
$$\therefore k = M \cdot \omega^2$$

Therefore the critical speed is: $\omega = k/Mr$.

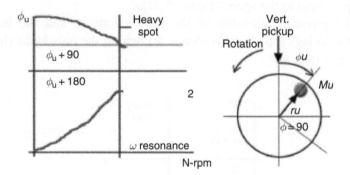

Figure 5.33
Phase relationship at 90°

Case 3 – running speed ω greater than the critical speed

As the speed increases beyond the critical speed, the mass stiffness contribution increases very rapidly (in quadratic proportion) and becomes higher in magnitude than the spring stiffness contribution, which almost remains the same in magnitude. The damping stiffness increases too but in linear proportion to the speed (Figure 5.34).

With the increase in synchronous dynamic stiffness, the amplitude of rotor vibration again drops, the phase difference continues to rise and by this stage it is close to 180°. This explains the nature of the Bode plot, which is a relationship of amplitude and phase vs running speed. The rising and falling of amplitude is due to the variations in the synchronous dynamic stiffness that changes at different speeds of the rotor (Figure 5.35).

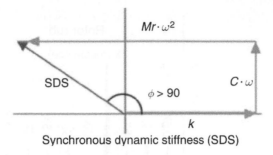

Figure 5.34
At critical speeds, mass stiffness rises above spring stiffness

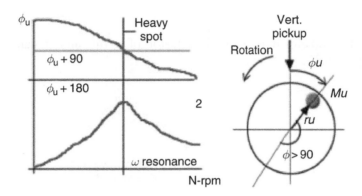

Figure 5.35
Dropping amplitude of rotor vibration

5.2.7 Rotor rubs

Rotor rubs produce a spectrum that is similar to mechanical looseness. A rub may be either partial or throughout the whole cycle. These generally generate a series of frequencies, and tend to excite one or more natural frequencies. Sometimes a phenomenon similar to chalk screeching on blackboard occurs during a rotor rub and it produces a white band noise on the spectrum in the high-frequency region. Normally the rub excites integer fractions of sub-harmonics of the running speed (½ , ⅓ , ¼ . . . 1/n), depending on the location of rotor natural frequencies.

The following relationships help to determine a rub. If N is the shaft speed and Nc is the critical speed of the shaft, then a rub will generate frequencies of:

$$1\times \quad \text{when } N < Nc$$
$$\tfrac{1}{2}\times \text{ or } 1\times \quad \text{when } N > 2Nc$$
$$\tfrac{1}{3}\times, \tfrac{1}{2}\times \text{ or } 1\times \quad \text{when } N > 3Nc$$
$$\tfrac{1}{4}\times, \tfrac{1}{3}\times, \tfrac{1}{2}\times \text{ or } 1\times \quad \text{when } N > 4Nc$$

A rub can have a short duration but can still be very serious if it is caused by the shaft touching a bearing (Figure 5.36). It is less serious when the shaft rubs a seal, an agitator blade rubs the wall of a vessel or a coupling guard presses against a shaft.

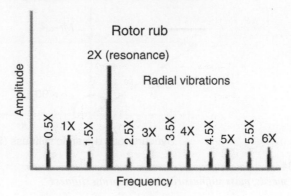

Figure 5.36
Rotor rub

The waveform is a good indicator of a rub and it can get truncated as shown in Figure 5.37.

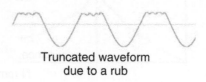

Figure 5.37
Truncated waveform

Orbit representation of a rotor rub

Orbit analysis is a good tool to identify rubs. As mentioned earlier, partial or complete rubs can occur when a rotating shaft comes in contact with stationary parts like seals or in abnormal cases of bearing (and/or instrumentation) failures. The rub causes the orbit to take on different shapes. From a number 8 to a full circle to something like the orbit shown in Figure 5.38.

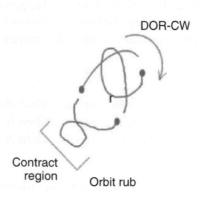

Figure 5.38
Orbit rub

A partial rub is more common than a complete or a full annular rub and occurs when the rotor occasionally touches a stationary part. This normally generates a ½× vibration. The orbit may then look like a number 8 (as seen under the topic of *misalignment under severe preloading*), except that we can see two dots on the orbit. A full circle orbit may be indicative of a complete rub in which the rotor fully covers the seal or bearing clearance. In this case, the precession of vibration is observed to be in reverse and must hence be rectified immediately.

5.2.8   Journal bearings

High clearance in journal bearings

Late stages of journal bearing wear normally display a whole series of running speed harmonics, which can be up to 10× or 20×. The FFT spectrum looks very much like that of mechanical looseness. Even minor unbalance or misalignment can cause higher vibration amplitudes compared with bearings having a normal clearance with the journal. This is due to a reduction in the oil film stiffness on account of higher clearances (Figure 5.39).

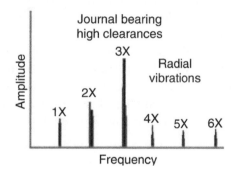

Figure 5.39
High clearance in journal bearings

Oil whirl

Oil whirl is an oil film-excited vibration. It is known to occur on machines equipped with pressure-lubricated journal bearings operating at high speeds (beyond their critical speed). Consider a shaft rotating in a bearing at speed *N*. The bearing speed is zero. The oil film is wedged between the shaft and the bearing and should ideally rotate at a speed of 0.5× rpm. However, some frictional losses cause the oil film to rotate at 0.42–0.48× rpm.

Under normal circumstances, the oil film pushes the rotor at an angle (5 o'clock if the shaft is rotating CCW – see Figure 5.40). An eccentric crescent-shaped wedge is created that has sufficient pressure to keep the rotor in the 'lifted' position. Under normal conditions, the system is in equilibrium and there are no vibrations.

Some conditions would tend to generate an oil film pressure in the wedge much higher than required to just hold the shaft. These conditions can cause an increase in bearing wear resulting in the shaft to have lower eccentricity (the shaft center is close to bearing center) causing a reduction in stiffness, oil pressure or a drop in oil temperature. In these cases, the oil film would push the rotor to another position in the shaft. The process continues over and over and the shaft keeps getting pushed around within the bearing. This phenomenon is called oil whirl. This whirl is inherently unstable since it increases centrifugal forces that will increase the whirl forces.

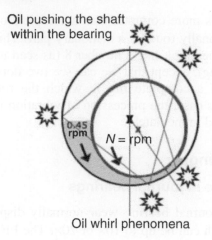

Figure 5.40
Oil whirl

Oil whirl can be minimized or eliminate4d by changing the oil velocity, lubrication pressure and external pre-loads. Oil whirl instability occurs at 0.42–048× rpm and is often quite severe. It is considered excessivewhen displacement amplitudes exceed 50% of the bearing clearances.

Oil whirl is basically a sub-synchronous fluid instability. When viewed in the orbit domain, it is shown with the characteristic two dots. When viewed with an oscilloscope, the two dots do not appear stationary, but seem to be rotating instead. This is because the frequency is marginally less than 0.5×. An oil whirl phenomenon generates a vibration precession, which is always forward (Figure 5.41).

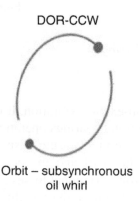

Figure 5.41
Orbit representation of an oil whirl

Oil whip

Oil whirl can be caused when the shaft has no oil support, and can become unstable when the whirl frequency coincides with a critical speed. This special coincidence of shaft resonance coupled with the oil whirl frequency results in a more severe form of oil whirl called oil whip. Whirl speed will actually 'lock' onto the rotor critical speed and will not disappear even if the machine is brought to higher and higher speeds (Figure 5.42).

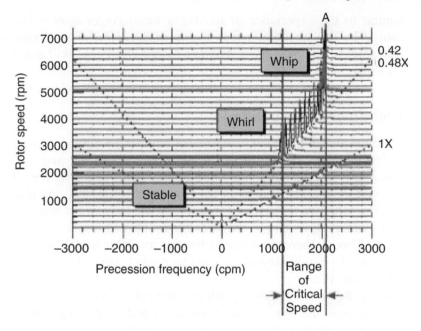

Figure 5.42
Oil whirl/whip as seen in a cascade full spectrum – note the oil whip frequency 'A' getting locked even after raising rotor speed

The oil whip phenomenon occurs when the rotor is passing through its critical speed. Oil whip is a destructive bearing defect. The precession of vibration is in the forward direction in this case, but some reverse 1× and sub-synchronous components are present due to anisotropy (changes in response when operating conditions change) of the bearing pedestal stiffness.

The period of this self-excited defect may, or might not, be harmonically related to the rotating speed of the shaft. When it is not harmonically related, the dots appear to be moving randomly as shown in Figure 5.43. When it is harmonically related they appear stationary.

DOR-CW

Orbit – subsynchronous
oil whip

Figure 5.43
Orbit representation of an oil whirl

Dry whirl

Sometimes inadequate or improper lubrication can also cause vibrations in a bearing. This is because lack of lubrication results in friction between the shaft and the bearing. The friction force will also tend to excite other parts of the machine. This vibration is similar to the experience of moving a moist finger over a glass pane. The vibration caused by this phenomenon is known as dry whirl. The vibration is generally at high frequencies, and harmonics may not be present. Phase will not provide any meaningful information.

5.2.9   Rolling element bearings

A rolling element bearing comprises of inner and outer races, a cage and rolling elements. Defects can occur in any of the parts of the bearing and will cause high-frequency vibrations. In fact, the severity of the wear keeps changing the vibration pattern. In most cases, it is possible to identify the component of the bearing that is defective due to the specific vibration frequencies that are excited. Raceways and rolling element defects are easily detected. However, the same cannot be said for the defects that crop up in bearing cages. Though there are many techniques available to detect where defects are occurring, there are no established techniques to predict when the bearing defect will turn into a functional failure.

In an earlier topic dealing with enveloping/demodulation, we saw how bearing defects generate both the bearing defect frequency and the ringing random vibrations that are the resonant frequencies of the bearing components.

Bearing defect frequencies are not integrally harmonic to running speed. However, the following formulas are used to determine bearing defect frequencies. There is also a bearing database available in the form of commercial software that readily provides the values upon entering the requisite bearing number.

$$BPFI = \frac{Nb}{2}(1 + \frac{Bd}{Pd}\cos\theta)\times\text{rpm}$$

$$BPFO = \frac{Nb}{2}(1 - \frac{Bd}{Pd}\cos\theta)\times\text{rpm}$$

$$FTF = \frac{1}{2}(1 - \frac{Bd}{Pd}\cos\theta)\times\text{rpm}$$

$$BSF = \frac{Pd}{2Bd}\left[1 - \left(\frac{Bd}{Pd}\right)^2(\cos\theta)^2\right]\times\text{rpm}$$

Nb = Number of Balls or Rollers
Bd = Ball / Roller diameter (inch or mm)
Pd = Bearing pitch diameter (inch or mm)
θ = Contact angle in degrees

BPFI = Ball pass frequency – Inner
BPFO = Ball pass frequency – Outer
FTF = Fundamental train frequency (Cage)
BSF = Ball spin frequency (rolling element)

It is very interesting to note that in an FFT, we find both the inner and outer race defect frequencies. Add these frequencies and then divide the result by the machine rpm – [(BPFI + BPFO)/rpm]. The answer should yield the number of rolling elements.

Bearing deterioration progresses through four stages. During the initial stage, it is just a high-frequency vibration, after which bearing resonance frequencies are observed. During the third stage, discrete frequencies can be seen, and in the final stage high-frequency random noise is observed, which keeps broadening and rising in average amplitude with increased fault severity.

Stage 1 of bearing defect

The FFT spectrum for bearing defects can be split into four zones (A, B, C and D), where we will note the changes as bearing wear progresses. These zones are described as:

Zone A: machine rpm and harmonics zone
Zone B: bearing defect frequencies zone (5–30 kcpm)
Zone C: bearing component natural frequencies zone (30–120 kcpm)
Zone D: high-frequency-detection (HFD) zone (beyond 120 kcpm).

The first indications of bearing wear show up in the ultrasonic frequency ranges from approximately 20–60 kHz (120–360 kcpm). These are frequencies that are evaluated by high-frequency detection techniques such as gSE (Spike Energy), SEE, PeakVue, SPM and others.

As Figure 5.44 shows, the raceways or rolling elements of the bearing do not have any visible defects during the first stage. The raceways may no longer have the shine of a new bearing and may appear dull gray.

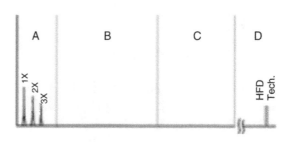

Figure 5.44
Small defects in the raceways of a bearing

Stage 2 of bearing defect

In the following stage (Figure 5.45), the fatigued raceways begin to develop minute pits. Rolling elements passing over these pits start to generate the ringing or the bearing component natural frequencies that predominantly occur in the 30–120 kcpm range. Depending on the severity, it is possible that the sideband frequencies (bearing defect frequency ± rpm) appear above and below the natural frequency peak at the end of stage two. The high-frequency detection (HFD) techniques may double in amplitude compared to the readings during stage one.

Stage 3 of bearing defect

As we enter the third stage (Figure 5.46), the discrete bearing frequencies and harmonics are visible in the FFT. These may appear with a number of sidebands. Wear is usually now visible on the bearing and may expand through to the edge of the bearing raceway.

The minute pits of the earlier stage are now developing into bigger pits and their numbers also increase. When well-formed sidebands accompany any bearing defect frequency or its harmonics, the HFD components have again almost doubled compared to stage three. It is usually advised to replace the bearing at this stage. Some studies indicate that after the third stage, the remaining bearing life can be 1 h to 1% of its average life.

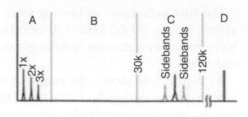

Figure 5.45
More obvious wear in the form of pits

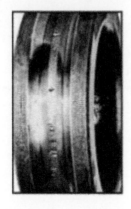
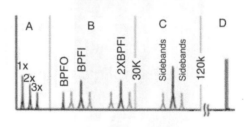

Figure 5.46
Wear is now clearly visible over the breadth of the bearing

Stage 4 of bearing defect

In the final phase (Figure 5.47), the pits merge with each other, creating rough tracks and spalling of the bearing raceways or/and rolling elements. The bearing is in a severely damaged condition now. Even the amplitude of the 1× rpm component will rise. As it grows, it may also cause growth of many running speed harmonics. It can be visualized as higher clearances in the bearings allowing a higher displacement of the rotor.

Discrete bearing defect frequencies and bearing component natural frequencies actually begin to merge into a random, broadband high-frequency 'noise floor'. Initially, the average amplitude of the broad noise may be large. However, it will drop and the width of the noise will increase. In the final stage, the amplitude will rise again and the span of the noise floor also increases.

 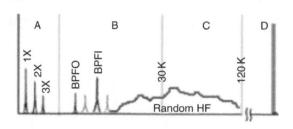

Figure 5.47
Severely damaged bearing in final stage of wear

However, amplitudes of the high-frequency noise floor and some of the HFD may in fact decrease (due to pits flattening to become spalls), but just prior to failure spike energy will usually grow to extreme amplitudes.

By this time, the bearing will be vibrating excessively; it will be hot and making lots of noise. If it is allowed to run further, the cage will break and the rolling elements will go loose. The elements may then run into each other, twisting, turning and welded to one another, until the machine will hopefully trip on overload. In all probability, there will be serious damage to the shaft area under the bearing.

5.2.10   Gearing defects

A gearbox is a piece of rotating equipment that can cause the normal low-frequency harmonics in the vibration spectrum, but also show a lot of activity in the high-frequency region due to gear teeth and bearing impacts. The spectrum of any gearbox shows the 1× and 2× rpm, along with the gear mesh frequency (GMF). The GMF is calculated by the product of the number of teeth of a pinion or a gear, and its respective running speed:

$$GMF = \text{number of teeth on pinion} \times \text{pinion rpm}$$

The GMF will have running speed sidebands relative to the shaft speed to which the gear is attached. Gearbox spectrums contain a range of frequencies due to the different GMFs and their harmonics. All peaks have low amplitudes and no natural gear frequencies are excited if the gearbox is still in a good condition. Sidebands around the GMF and its harmonics are quite common. These contain information about gearbox faults (Figure 5.48).

Tooth wear and backlash can excite gear natural frequencies along with the gear mesh frequencies and their sidebands. Signal enhancement analysis enables the collection of vibrations from a single shaft inside a gearbox.

Cepstrum analysis is an excellent tool for analysing the power in each sideband family. The use of cepstrum analysis in conjunction with order analysis and time domain averaging can eliminate the 'smearing' of the many frequency components due to small speed variations (Figure 5.49).

As a general rule, distributed faults such as eccentricity and gear misalignment will produce sidebands and harmonics that have high amplitude close to the tooth-mesh frequency. Localized faults such as a cracked tooth produce sidebands that are spread more widely across the spectrum.

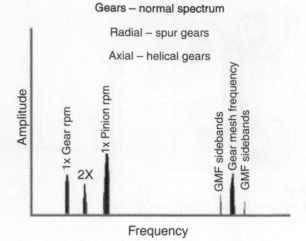

Figure 5.48
Graph of a gearbox spectrum

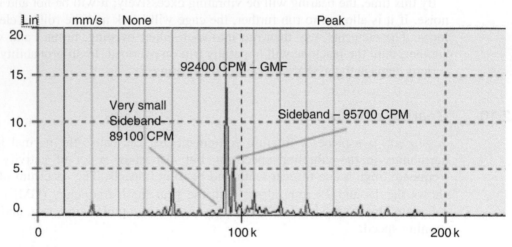

Figure 5.49
FFT spectrum from a noisy gearbox with pinion having 28 teeth and rotating at 3300 rpm

Gear tooth wear

An important characteristic of gear tooth wear is that gear natural frequencies are excited with sidebands around them. These are spaced with the running speed of the bad gear. The GMF may or may not change in amplitude, although high-amplitude sidebands surrounding the GMF usually occur when wear is present. Sidebands are a better wear indicator than the GMF itself (Figure 5.50).

Gear tooth load

As the load on a gearbox increases, the GMF amplitude may also increase. High GMF amplitudes do not necessarily indicate a problem, particularly if sideband frequencies remain low and no gear natural frequencies are excited. It is advised that vibration analysis on a gearbox be conducted when the gearbox is transmitting maximum power (Figure 5.51).

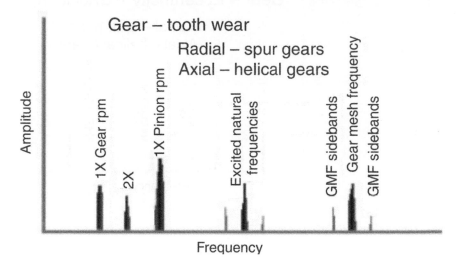

Figure 5.50
Gear tooth wear

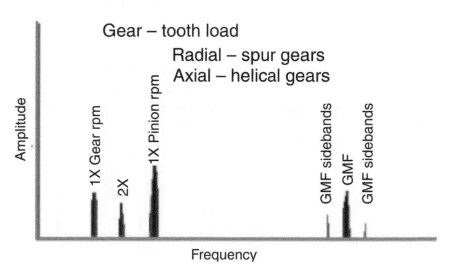

Figure 5.51
Gear tooth load

Gear eccentricity and backlash

Fairly high amplitude sidebands around the GMF often suggest gear eccentricity, backlash or non-parallel shafts. In these cases, the rotation of one gear may cause the amplitude of gear vibration to *modulate* at the running speed of the other. This can be seen in the time domain waveform. The spacing of the sideband frequencies indicates the gear with the problem. Improper backlash normally excites the GMF and gear natural frequencies. Both will have sidebands at 1× rpm. The GMF amplitudes will often decrease with increasing load if backlash is the problem (Figure 5.52).

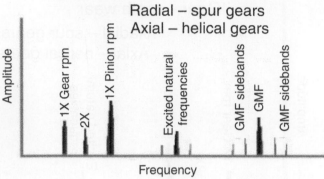

Figure 5.52
Gear eccentricity and backlash

Gear misalignment

Gear misalignment almost always excites second order or higher GMF harmonics, which will have sidebands spaced with the running speed. It will often show only small amplitudes at 1× GMF, but much higher levels at 2× or 3× GMF. It is important to set the F-max of the FFT spectrum to more than 3× GMF (Figure 5.53).

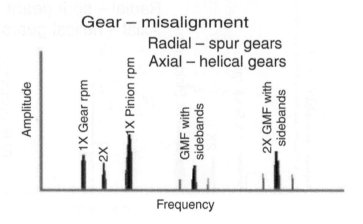

Figure 5.53
Gear misalignment

Gears – cracked or broken tooth

A cracked or broken gear tooth will generate high amplitude at 1× rpm of this gear, plus it will excite the gear natural frequency with sidebands spaced with its running speed. It is best detected in the time domain, which will show a pronounced spike every time the problematic tooth tries to mesh with teeth on the mating gear. The time between impacts will correspond to 1/speed of the gear with the broken tooth. The amplitude the impact spike in the time waveform will often be much higher than that of the 1× gear rpm in the FFT spectrum (Figure 5.54).

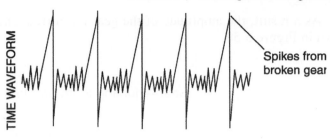

Figure 5.54
Gears – cracked

Gears – hunting tooth problems

The gear hunting tooth frequency is particularly effective for detecting faults on both the gear and the pinion that might have occurred during the manufacturing process or due to mishandling. It can cause quite high vibrations, but since it occurs at low frequencies, predominantly less than 600 cpm, it is often missed during vibration analysis. The hunting tooth frequency is calculated with:

$$\text{Hunting Tooth Frequency} = \frac{\text{GMF} \times \text{N}}{(\text{no. of pinion teeth}) \times (\text{no. of gear teeth})}$$

In the above equation, *N* is known as the *assembly phase factor*, also referred to as the lowest common integer multiple between the number of teeth on the pinion and gear. This hunting tooth frequency is usually very low.

For assembly phase factors (*N* > 1), every gear tooth will not mesh with every pinion tooth. If *N* = 3, teeth numbers 1, 4, 7, etc. will mesh with one another (however, gear tooth 1 will not mesh with pinion teeth 2 or 3; instead, it will mesh with 1, 4, 7, etc.). For example, a gear with 98 teeth is running at 5528 rpm and is meshing with a pinion with 65 teeth and running 8334 rpm. The assembly phase factor is *N* = 1. The hunting tooth frequency (Fht) can be calculated as follows:

$$\text{Fht} = \frac{(98 \times 5528) \times 1}{98 \times 65} = 85 \text{ cpm (1.42 Hz)}$$

Another formula is the rpm of the gear divided by the number of pinion teeth (5528/65 = 85 cpm). This is a special case and applies to a hunting tooth combination only when *N* = 1.

If the tooth repeat frequency is a problem (Figure 5.55), one can usually audibly hear it since it is a beat frequency. A gearset with a tooth repeat problem normally emits a 'growling' sound from the driven end. Its repetition rate can often be established by simply counting the sounds using a stopwatch. The maximum effect occurs when the faulty pinion and gear teeth mesh at the same time (on some drives, this may occur once every 10 or 20 revolutions, depending on the Fht formula).

Gearboxes can generate crowded and complex FFT spectrums with many unusual and unidentifiable frequencies. Another unusual frequency encountered in the gearbox is a sub-multiple of the gear mesh frequency. This vibration is generally the result of an eccentric gear shaft misalignment or possibly a bent shaft. Any one of these cases can cause variations in the tooth clearances for each revolution of the gear. As a result, the amplitude of the gear mesh frequency may appear modulated as shown in Figure 5.56.

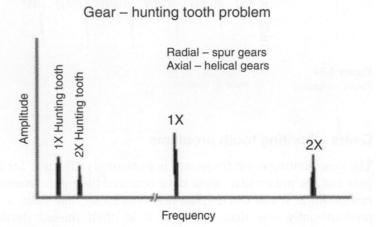

Figure 5.55
Gear – hunting tooth problem

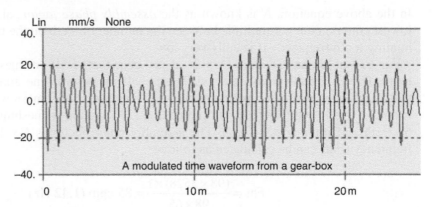

Figure 5.56
Modulated amplitude from a gearbox with a GMF of 92400 cpm

Gear tooth impacts may only be excessive during portions of each revolution of the gear. Therefore the resulting vibration will have a frequency less than the gear mesh frequency. However, it will remain a multiple of the gear rpm.

5.2.11   Belt defects

Worn, loose, mismatched belts

Belt defect frequencies are of the sub-harmonic type. Upon analysing belt drives, it is necessary to keep the F-max low to be able to notice these peaks. When belts are worn, loose or mismatched, they may generate harmonics of the belt frequency. It is possible to obtain 3× or 4× times of belt frequency. Quite often, the 2× belt frequency is dominant. The belt frequency (Figure 5.57) is given by the formula:

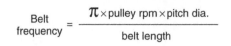

$$\frac{\text{Belt}}{\text{frequency}} = \frac{\pi \times \text{pulley rpm} \times \text{pitch dia.}}{\text{belt length}}$$

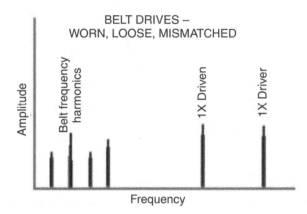

Figure 5.57
Sub-harmonic belt frequencies

Amplitudes are normally unsteady, sometimes pulsing with either driver or driven rpm. With timing belt drives, it is useful to know that high amplitudes at the timing belt frequency indicate wear or pulley misalignment.

Belt/sheave misalignment

The different types of misalignment possible with belt drives are shown in Figure 5.58. These conditions not only result in destructive vibration but also cause accelerated wear of both the belt and the sheaves. Misalignment of sheaves produces high vibration at 1× rpm, predominantly in the axial direction (Figure 5.59). The ratio of amplitudes of driver to driven rpm depends on the measurement position, relative mass and the frame stiffness. With sheave misalignment in fans, the highest axial vibration will be at the fan rpm. When the belt drives an overhung rotor, which is in an unbalanced condition, it will have to be resolved with phase analysis.

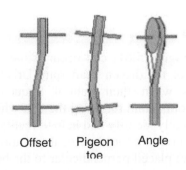

Figure 5.58
Misalignment types (the pigeon toe and angle are classified as angular misalignment)

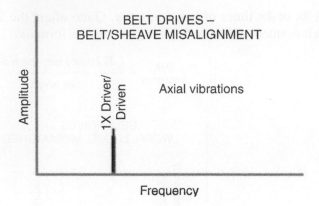

Figure 5.59
Vibration due to sheave misalignment

Eccentric sheaves

Eccentric or unbalanced sheaves cause maximum vibration at 1× rpm of the sheave, causing problems in line with the sheaves. To resolve this condition, it may sometimes be possible to balance eccentric sheaves by attaching washers to taperlock bolts. However, even if balanced, the eccentricity will still induce vibration and cause fatigue stresses in the belt (Figure 5.60).

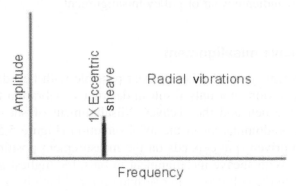

Figure 5.60
Belt drives – eccentric sheaves

Belt resonance

Belt resonance (Figure 5.61) can occur if the natural frequency of the belt is close to either the motor or the driven shaft rpm. Drive belts also experience high vertical and lateral vibrations when their natural frequencies coincide with that of connected equipment. Tensioning and releasing the belt while measuring the response on sheaves or bearings can normally identify and help to rectify this situation.

A common method to control vertical vibration is by using a restraining device (metal rod or idler pulley) placed perpendicular to the belt span and close to (or lightly touching) the belt. This device should be positioned roughly at one-third of the span distance from the larger pulley.

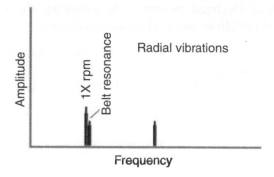

Figure 5.61
Belt drives – resonance

Other alternatives to reduce the amplitude of vibration or alter the belt resonance frequency are to modify the span length, belt type, alignment, inertia of driving or driven machinery, pulley diameter and weight, speed, and the number of belts.

5.2.12 Electrical problems

Vibrations of electrical machines such as motors, generators and alternators can be either mechanical or electrical in nature. We have discussed most common mechanical problems. Electrical problems also appear in the vibration spectrum and can provide information about the nature of the defects. Electrical problems occur due to unequal magnetic forces acting on the rotor or the stator. These unequal magnetic forces may be due to:

- Open or short windings of rotor or stator
- Broken rotor bar
- Unbalanced phases
- Unequal air gaps.

Generally, the vibration pattern emerging due to the above-mentioned electrical problems will be at 1× rpm and will thus appear similar to unbalance.

A customary technique to identify these conditions is to keep the analyzer capturing the FFT spectrum in the *live* mode and then switching off the electrical power. If the peak disappears instantly, the source is electrical in nature. On the other hand, if there is gradual decrease in the 1× amplitude it is more likely to be a mechanical problem. This technique requires caution. If there is a time lag in the analyzer itself, it may delay the drop in vibration amplitude. It is also possible that a resonance frequency may drop quickly as the speed changes.

Induction motors, which have electrical problems, will cause the vibration amplitude to hunt or swing in a cyclic manner. The phase readings will show similar cycles too. Under a stroboscope, the reference mark will move back and forth.

The swinging amplitudes in induction motor applications are due to two dominant frequencies that are very close to one another. They continuously add and subtract to one another in a phenomenon known as *beats*. It can also possibly be a single frequency whose amplitude is modulating.

In fact, hunting amplitudes are the first indication of a possible electrical problem in the motor. Understanding the nature of these vibrations can assist in identifying the exact defects in an electrical machine. The following are some terms that will be required to understand vibrations due to electrical problems:

$$F_L = \text{electrical line frequency (50/60 Hz)}$$

$$F_s = \text{slip frequency} = \frac{2 \times F_L}{P} - \text{rpm}$$

$$F_p = \text{pole pass frequency} = F_s \times P$$

$$P = \text{number of poles.}$$

Rotor problems

Normally, four kinds of problems can occur within the rotor:

1. Broken rotor bars
2. Open or shorted rotor windings
3. Bowed rotor
4. Eccentric rotor.

Rotor defects

Along with the stator is a rotor, which is basically an iron following the rotating magnetic field. As the magnetic field sweeps across the conductor, it creates a voltage across the length of the rotor bar. If the bar is open-circuited, no current flows and no forces are generated. When the bar is short-circuited, a current flows. This current is proportional to the speed at which the field cuts through the conductor and the strength of the field. The field interacts with the stator field to generate a force on the rotor bar. If everything else remains the same, an equal and opposite force on the opposite side of the rotor will develop. These two forces generate the torque that drives the load. In case anything disrupts the current or magnetic fields on either side of the rotor, the two forces will become unequal. This results in a radial force, which is the cause for vibration.

A cracked or broken bar can cause this category of unbalanced forces. The forces rotate with the rotor with a constant load plus a load that varies with 2× slip. Therefore, the force acting on the bearings will have frequency components at 1× rpm and 1× rpm ± 2× slip. Thus:

- Broken or cracked rotor bars or shorting rings (Figures 5.62 and 5.63)
- Bad joints between rotor bars and shorting rings
- Shorted rotor laminations

will produce high 1× running speed vibration with pole pass frequency sidebands. In addition, cracked rotor bars will often generate F_P sidebands around the 3rd, 4th and 5th running speed harmonics.

Loose rotor bars are indicated by 2× line frequency ($2F_L$) sidebands surrounding the rotor bar pass frequency (RBPF) and/or its harmonics (Figure 5.64).

$$\text{RBPF} = \text{number of rotor bars} \times \text{rpm}$$

It may often cause high levels at 2× RBPF with only a small amplitude at 1× RBPF.

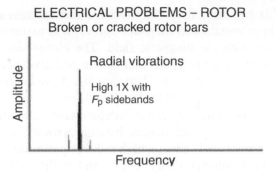

Figure 5.62
High 1✕ with F_P sidebands

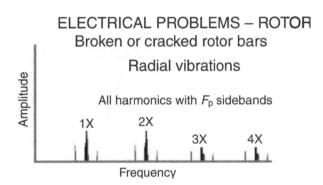

Figure 5.63
All harmonics with F_P sidebands

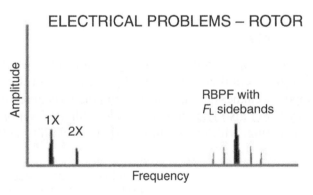

Figure 5.64
Rotor bar pass frequency

Eccentric rotor

The rotor is supposed to be concentric with respect to the stator coils. If this is not the case, a magnetic unbalance force is generated which is given by the formula:

$$F = \frac{KI^2}{g^2}\left(\frac{4\,e}{(1-e)^2}\right)$$

where I = stator current, g = average gap between stator and rotor, e = eccentricity.

From this equation, it can be observed that an increase in current and eccentricity can generate high unbalanced magnetic forces. It is assumed that the eccentricity of the rotor will line up with the magnetic field. The closer side of the rotor will be respectively attracted to the positive pole and to the negative pole; thus the force will vary twice during a single current cycle. This can affect the bearings, and therefore it can modulate any other frequency present in the system.

These effects generally cause sidebands of $\pm2\times$ slip frequency around the $1\times$ rpm frequency caused by unbalance. Eccentric rotors produce a rotating variable air gap between the rotor and stator, which induces pulsating vibrations (it is a beat phenomenon between two frequencies, one is $2F_L$ and is the closest running speed harmonic). This may require a 'zoom' spectrum to separate the $2F_L$ and the running speed harmonic.

Eccentric rotors generate $2F_L$ surrounded by pole pass frequency sidebands (F_P as well as F_P sidebands around $1\times$ rpm). The pole pass frequency F_P itself appears at a low frequency (Figure 5.65).

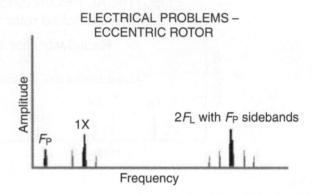

Figure 5.65
Eccentric rotor

Stator defects

An induction motor comprises a set of stator coils, which generate a rotating magnetic field. The magnetic field causes alternating forces in the stator. If there is any looseness or a support weakness in the stator, each pole pass gives it a tug. This generates a $2\times$ line frequency ($2F_L$) also known as *loose iron*. Shorted stator laminations cause uneven and localized heating, which can significantly grow with time (Figure 5.66).

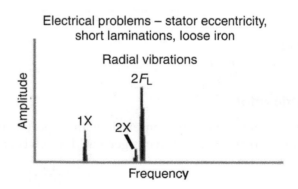

Figure 5.66
Stator defects

Stator problems generate high vibration at $2F_L$. Eccentricity produces uneven stationary air gaps between the rotor and the stator, which produce very directional vibration. Differential air gaps should not exceed 5% for induction motors and 10% for synchronous motors. Soft foot and warped bases can generate an eccentric stator.

Phasing problem (loose connector)

Phasing problems due to loose or broken connectors can cause excessive vibration at $2F_L$, which will have sidebands around it spaced at $\frac{1}{3}$ rd of the line frequency ($\frac{1}{3}$ F_L). Levels at $2F_L$ can exceed 25 mm/s (1.0 in./s) if left uncorrected. This is particularly a problem if the defective connector is sporadically making contact and not periodically (Figure 5.67).

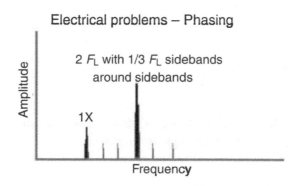

Figure 5.67
Phasing problem

Synchronous motors (loose stator coils)

Loose stator coils in synchronous motors will generate fairly high vibrations at the coil pass frequency (CPF), defined as:

$$CPF = \text{number of stator coils} \times \text{rpm}$$

(number of stator coils = poles × number of coils/pole)
The coil pass frequency will be surrounded by 1× rpm sidebands (Figure 5.68).

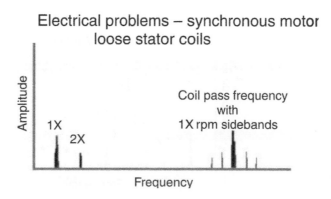

Figure 5.68
Synchronous motors

DC motor problems

DC motor defects can be detected by high vibration amplitudes at the SCR firing frequency ($6F_L$) and harmonics. These defects include broken field windings, bad SCRs and loose connections (Figure 5.69).

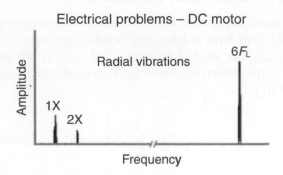

Figure 5.69
DC Motor

Other defects, such as loose or blown fuses and shorted control cards, can cause high amplitude peaks at 1× through 5× line frequency.

5.2.13   Flow-related vibrations

Blade pass and vane pass vibrations

Blade pass or vane pass frequencies (Figure 5.70) are characteristics of pumps and fans. Usually it is not destructive in itself, but can generate a lot of noise and vibration that can be the source of bearing failure and wear of machine components.

$$\text{Blade pass frequency (BPF)} = \text{number of blades (or vanes)} \times \text{rpm}$$

This frequency is generated mainly due to the gap problems between the rotor and the stator. A large amplitude BPF (and its harmonics) can be generated in the pump if the gap between the rotating vanes and the stationary diffusers is not kept equal all the way around.

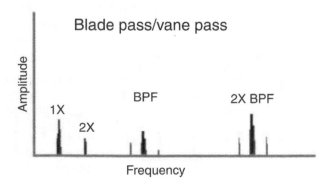

Figure 5.70
Blade pass/vane pass

In centrifugal pumps, the gap between the impeller tip and the volute tongue or the diffuser inlet is a certain percentage (in the region of 4–6% of the impeller diameter), depending on the speed of the pump. If the gap is less than the recommended value, it can generate a noise that resembles cavitation. However, an FFT plot will immediately highlight the vane pass frequency of the impeller. Also, the BPF (or its harmonics) sometimes coincides with a system natural frequency, causing high vibrations.

A high BPF can be generated if the wear ring seizes on the shaft or if the welds that fasten the diffusers fail. In addition, a high BPF can be caused by abrupt bends in linework (or duct), obstructions which disturb the flow path, or if the pump or fan rotor is positioned eccentrically within the housing.

Similar to the vane pass frequency, centrifugal pumps are known to generate non-specific sub-synchronous or even supersynchronous (larger than 1×) discrete frequencies. These are rare occurrences, but in all probability they transpire in two-stage (or higher) pumps, which have intermediate bushes that act as additional stiffness components. An increase in the clearances within these bushes leads to a fall in stiffness and this results in enlarged vibrations.

In a two-stage overhung impeller pump, the interstage bushing plays an important role in providing stiffness, which is described as the Lomakin effect. When clearances are high, this effect can reduce and high amplitude supersynchronous frequencies are generated. Once the clearances are adjusted back to normal, the pump operation stabilizes and the defect frequency disappears.

Flow turbulence

Flow turbulence (Figure 5.71) often occurs in blowers due to variations in pressure or velocity of the air passing through the fan or connected linework. In fans, duct-induced vibration due to stack length, ductwork turns, unusual fan inlet configuration and other factors may be a source of low-frequency excitation. This flow disruption causes turbulence, which will generate random, low-frequency vibrations, typically in the range of 20–2000 cpm.

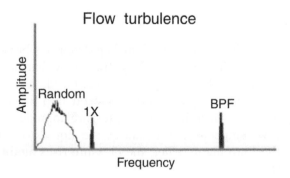

Figure 5.71
Flow turbulence

Rotating stall is one of the flow-induced vibrations that can occur in fans and compressors. Rotating stall is a flow separation of the fluid from the blades under certain low-flow conditions. Rotating stall sometimes occur in a system with a partially closed inlet damper. The condition usually appears as a low sub-synchronous frequency component in the rotor vibration spectrum (frequency ratios are typically between 8 and 40%, but can be as high as 80% of the rotational speed).

From a diagnostics point of view, rotating stall differs from the other whirl category instabilities due to its strong dependence on the *operating conditions*. Normally, correcting the operating flow makes it disappear. It differs from surge because it is proportional to the running speed of the fan or compressor. Surging is in the axial direction, which is not the case with rotating stall. Rotating stall manifests in the rotor vibration spectrum with sub-synchronous frequencies, which tracks the rotor speed. The orbit will have a forward precession.

In pumps, flow turbulence induces vortices and wakes in the clearance space between the impeller vane tips and the diffuser or volute lips. Dynamic pressure fluctuations or pulsation produced in this way can result in shaft vibrations because the pressure pulses impinge on the impeller.

Flow past a restriction in pipe can produce turbulence or flow-induced vibrations. The pulsation could produce noise and vibration over a wide frequency range. The frequencies are related to the flow velocity and geometry of the obstruction. These in turn excite resonant frequencies in other pipe components. The shearing action produces vortices that are converted to pressure disturbances at the pipe wall, which may cause localized vibration excitation of the pipe or its components.

It has been observed that vortex flow is even higher when a system's acoustic resonance coincides with the generated frequency from the source. The vortices produce broadband turbulent energy centered around the frequency determined by the following formula:

$$f = \frac{S_n \times V}{D}$$

where f = vortex frequency (Hz), S_n = Strouhl number (dimensionless, between 0.2 and 0.5), D = characteristic dimension of the obstruction.

For example, a liquid flowing at 35 m/s past a 6-in. stub line would produce broadband turbulence at frequencies from 40 to 100 Hz.

Cavitation

Cavitation normally generates random, high-frequency broadband energy, which is sometimes superimposed with the blade pass frequency harmonics. Gases under pressure can dissolve in a liquid. When the pressure is reduced, they bubble out of the liquid.

In a similar way, when liquid is sucked into a pump, the liquid's pressure drops. Under conditions when the reduced pressure approaches the vapor pressure of the liquid (even at low temperatures), it causes the liquid to vaporize. As these vapor bubbles travel further into the impeller, the pressure rises again causing the bubbles to collapse or implode.

This implosion has the potential to disturb the pump performance and cause damage to the pump's internal components. This phenomenon is called *cavitation*. Each implosion of a bubble generates a kind of impact, which tends to generate high-frequency random vibrations (Figure 5.72).

Cavitation can be quite destructive to internal pump components if left uncorrected. It is often responsible for the erosion of impeller vanes. Cavitation often sounds like 'gravel' passing through the pump. Measurements to detect cavitation are usually not taken on bearing housings, but rather on the suction piping or pump casing.

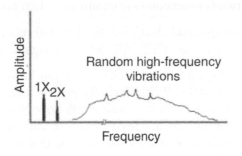

Figure 5.72
Cavitation

Cavitation is best recognized by observing the complex wave or dynamic pressure variation using an oscilloscope and a pressure transducer. The pressure waveform is not sinusoidal, and the maximum amplitudes appear as sharp spikes. Between these spikes are low amplitude, smooth and rounded peaks.

5.2.14   Rotor crack

The basic principle during crack development is that the rotor loses stiffness in the direction perpendicular to the crack direction. Imagine a flat steel ruler. Tie a heavy weight to one end of the ruler with the help of a string. As we turn the ruler, we see a big deflection when the broad and flat surface is on top. When it is turned through 90°, the thin section of the ruler is on top and this time we hardly notice any deflection.

Thus, in one revolution of the ruler we will see two big deflections, and in two instances there will be almost zero deflection. The two big deflections per revolution would cause the 2× rpm vibration frequency.

This same principle applies to a shaft under a heavy side load, such as a turbine rotor acting under gravity. If a crack develops on the circumference of a rotor, transverse to the shaft axis, the stiffness in the plane perpendicular to the crack decreases and remains the same in the other orthogonal plane (Figure 5.73).

This is similar to the flat steel ruler example, and therefore we will observe an analogous phenomenon. To diagnose this fault properly, we must look at all the information obtained from the vibration amplitude and the phase data carefully.

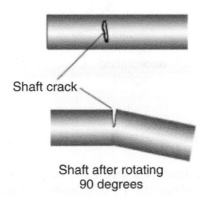

Figure 5.73
Shaft crack

There are two fundamental symptoms of a shaft crack:

1. Unexplained changes in the 1× shaft relative amplitude and phase
2. Occurrence of a 2× rpm vibration frequency.

The first vital symptom is the changes in the 1× synchronous amplitude and phase. On a turbomachine installed with proximity probes measuring shaft vibration, it may also be possible to notice the defect on a slow roll bow vector (at a low rotational speed).

Shaft bending due to a transverse crack causes the changes in the synchronous 1× amplitude and phase. It is important to note that the amplitude and phase could be higher or even lower. Thus, any change in the 1× amplitude and phase should cause an alarm for the possibility of this defect.

The next classical symptom is the occurrence of the 2× rpm component. The cause for this component (as explained earlier) is due to asymmetry in the horizontal shaft stiffness on which radial forces like gravity may be acting. When viewed on a cascade plot, during a startup or coastdown, the 2× frequency is especially dominant when the rotational speed is in the region of half the critical speed.

When the speed is increased, this 2× peak diminishes. When the rotor crosses the critical speed, the 2× amplitude also rises along with 3× and 4× components that may also be present. When the rotor reaches full speed, the high 1× may be accompanied by this 2× frequency too.

Although this is a transient observation, the 1× amplitude and phase can be monitored under normal operating conditions as well to provide alarms and early warnings of a possible shaft crack. A polar plot (Figure 5.74) provides a good format for highlighting the change in the 1× amplitude and phase.

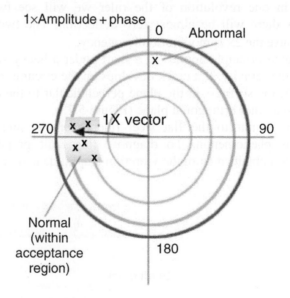

Figure 5.74
Polar plot

A region in the form of a sector can be plotted to indicate the normal operating vector position that describes the 1× amplitude and phase. These are also called *acceptance regions*. Acceptance regions can also be plotted for the 2× component during the transient analysis to provide evidence of a shaft crack.

Any change in the position of this vector will cause it to move away from the sector. Alarm windows can be created around this sector to draw attention to a deviation.

It should be kept in mind that many other factors such as load, field current, steam conditions or other operating parameters could have changed and might be causing the changes in the 1× and 2× amplitude and phase readings. A thermal blow in a large steam turbine can cause a similar high 1× component. Misalignment can cause large 1× and 2× components. In some cases, the high 1× amplitude was associated with unbalance. However, if a shaft cannot be balanced properly, a crack could be the culprit.

6

Correcting faults that cause vibration

6.1   Introduction

No predictive maintenance program is complete until it has the three basic components: detection, analysis and correction. In the previous topics, we discussed vibration detection and analysis of machinery faults in detail. Statistics indicates that a very large percentage of machinery vibrations are due to unbalance and misalignment alone. In this section, we will discuss the correction of these common faults.

The balancing of rotors is possible in the field, but can also be done with dedicated machines. In this section, we will discuss these techniques.

Similarly, misalignment is also a major cause of unwanted vibration. Alignment correction also requires special techniques. The latest techniques that have simplified these procedures are also covered in this section.

When excessive vibrations due to resonance are encountered, it is often difficult to find an easy solution to the problem. The use of *dynamic absorbers* as a possible tool for controlling resonance-induced vibration will also be discussed.

6.2   Balancing

It is very rare to come across a vibration spectrum of a machine that does not show a $1\times$ component indicating some residual unbalance. Unbalance causes high levels of vibration amplitudes at the $1\times$ rpm of the machine.

Unbalance is often defined as simply the unequal distribution of weight about a rotor's centerline. The ISO defines it as a condition that exists in a rotor when the vibratory force or motion is imparted to its bearings as a result of centrifugal forces. Correction of this unequal distribution of weight about a rotor is called *balancing*. Balancing compensates for less-than-perfect manufacturing. The main causes for unbalance during manufacture are:

- Materials do not have uniform density
- Holes are not bored exactly parallel to the center
- Imperfections occur in machining round or symmetrical shapes
- Assembly errors.

Unbalance can also occur during normal operation of machines. These causes are listed below:

- Uneven product deposition around a fan or pump impeller
- Damaged or missing blades or vanes
- Thermal distortion of the rotor due to temperature deviations in the process.

6.2.1   Balancing concepts

To improve an unbalanced rotor, it is necessary to determine the location and weight of the *heavy spot*. The heavy spot is the radial location where the excessive radial mass distribution exists. Attaching an equal mass in an opposite location can improve the effect of this excessive distribution. The dilemma is that it is difficult to identify the location of the heavy spot. However, it is possible to identify the *high spot*. The high spot is the radial location where the shaft experiences maximum displacement during a rotation.

There is a deterministic correlation between the heavy spot and the high spot. For rotors that run at speeds below the critical speed, the heavy spot and the high spot are in the same location. However, we have seen in an earlier section on resonance that the heavy spot and the high spot can be as much as 180° apart after crossing the critical speed.

If the rotor continues to increase in speed and passes through another critical speed, the high spot rotates another 180° until the high spot coincides with the heavy spot again. This phase shift between the high spot and the heavy spot continues as subsequent critical speeds are passed. Rectifying an unbalance condition involves phase measurements to locate the high spot, determining the relationship of the high spot to the heavy spot, and finding the magnitude of the unbalance by measuring the influence of correction weights.

In order to determine the unbalance, it is essential to learn how a perfectly balanced rotor responds to a mass that would tend to make it unbalanced. When an unbalance weight is added to a perfectly balanced rotor:

- It vibrates at a frequency of 1× rpm.
- The measured phase will be steady, and under the strobe light a reference mark will appear stationary (at any angle).
- If the weight of the unbalance is doubled, the vibration amplitude also doubles, which implies the amplitude is proportional to the unbalance force. As an example, we attach an unbalance weight of 30 g to a perfect rotor and observe that the vibration amplitude is 2. After this, the 30-g weight is removed and a 60-g weight is attached at the same location. The vibration amplitude will be 4. This is an extremely important fact to note: *The vibration amplitude is an indicator of the severity of unbalance.*
- If the location of the unbalance weight is changed, the phase reading will change. Under the stroboscope, the reference mark will appear in a different location. *The phase is an indicator of the location of the unbalance.*
- If the unbalance weight is moved clockwise through a certain number of degrees, the phase reading or the reference mark under the strobe light will move by an equal number but in the *opposite* direction. Consider the previous example again. Let us say the reference mark under the strobe light appeared at 90° with the 30-g unbalance weight. Subsequently, the 30 g is moved through 50° in a clockwise direction. Now we will notice that the reference mark also moved through 50° but in a counter-clockwise direction, to indicate 40°.

Hence, the two important fundamentals of balancing are:

1. The amplitude of vibration is proportional to the severity of unbalance.
2. Reference marks shift in the direction opposite to the heavy spot. However, the angle of the shift of the heavy spot and the reference mark is the same.

6.2.2   The effect of a trial weight

We now consider a realistic case, where we do not start with a perfectly balanced rotor, but with a rotor that is intrinsically unbalanced. To begin with, we have no idea of the location or the severity of unbalance in the rotor. The first step involves measuring the vibration and phase of the rotor. Upon taking a trial measurement we notice that the amplitude of vibration is 5 and the phase angle is 120°. At this stage, the above data is not complete enough to provide us with the information required to determine the unbalance in terms of its severity and location.

To get the complete picture of the original unbalance, the rotor has to be disturbed to determine the rotor's response due to the disturbance. This is achieved by using a trial weight. A trial weight is a mass of known weight that is attached at a specific location with respect to the reference mark. When a trial weight is added, the original unbalance is disturbed. It may change for the better or worse, or not at all. If the original unbalance does not change at all, it implies that the selection of the trial weight or its location on the rotor is inadequate. A heavier trial weight must be attached, or a different axial location must be used. As a rule of thumb, a trial weight should result in at least 30% change in vibration amplitude and phase.

The change in unbalance provides us with another set of vibration and phase readings. In our example, we now attach a 10-g mass as a trial weight on the rotor. The result is vibration and phase readings of 8 and 30°. After we have obtained this information, it is possible to locate the magnitude and location of the heavy spot by using the vector method.

6.2.3   Balancing methods

Single plane balancing – vector method

Single plane balancing is used for machines that operate below their critical speed and have an *L/D* ratio lower than 0.5 (*L*, length of rotor, excluding support length; *D*, diameter of the rotor). It is also recommended to avoid the use of this method for rotors operating at speeds greater than 1000 rpm. For cases where *L/D* ratio is greater than 0.5 but less than 2, this method should be applied for rotors that do not operate beyond 150 rpm. For *L/D* ratios greater than 2, the limit is 100 rpm. For this method of balancing, the following steps are taken:

- During the first run, the original vibration and phase readings are recorded. For example, if the readings obtained are 6 mils (or mm, micron, etc.) at a phase angle of 60°, a polar plot can be constructed and a vector proportional to 6 mils is drawn at an angle of 60° from the 0° reference. This vector is called O (Figure 6.1).
- In the next step, a trial weight of 20 g is attached to the rotor at any location. During the subsequent measurement, we obtain the vibration and phase readings of 4 mils at 150°. It should be noted that the new values are due to the combined effect of the original unbalance and that of the trial weight. Let us call this vector O + T (Figure 6.2).

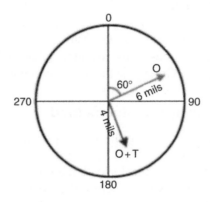

Figure 6.1
O vector

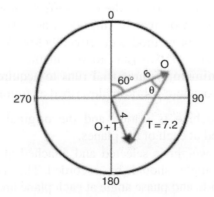

Figure 6.2
O + T vector

- The next step is to join the vectors O and O + T. The resultant vector is a result of the trial weight and hence should be designated as vector T (Figure 6.3).
- The vector T is measured and scaled, and found to be 7.2 mils in magnitude. The angle between the vector O and vector T called θ is 33.7°.

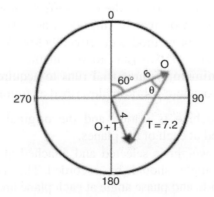

Figure 6.3
Vector T

With the above results we are in a position to determine the correction required to nullify the original unbalance of the rotor. The question is, if a vector T of 7.2 mils is generated by a trial weight of 20 g, what is the weight in grams that caused the original vector O?

$$\text{correct weight} = \frac{\text{TW} \times \text{vector O}}{\text{vector T}}$$
$$= \frac{20 \times 6}{7.2}$$
$$= 16.7\,\text{g}$$

Mathematically, we need to move the vector T in such a way to cancel the vector O. The movement should be equal in magnitude and opposite in direction. The angle between vector O and vector T is measured as 33.7°. Here, the vector T has to be moved clockwise to make it opposite to vector O.

The new weight should be 16.7 g. It should be moved through 33.7° in the counter-clockwise direction (due to *fundamental two* of balancing) from its original point. Small errors due to angle measurement, positioning and other factors will result in a small residual unbalance. In case this residual unbalance is within the specified limits, the balancing is complete. Otherwise the above procedure must be repeated.

Two plane balancing – vector method

Single plane balancing has very limited application. Generally, all machines that have an *L/D* ratio of more than 0.5 should be balanced using the two-plane method. However, if a machine runs above it in critical speed, then the rule is $N + 2$ planes for balancing, where N is number of critical speeds below the operating speed. For example, a compressor operating above its first critical speed should be balanced using the three-plane balancing method.

Two-plane balancing is done in a manner similar to the single plane method. Two-plane balancing requires special attention due to the cross-effect or correction plane interference. Cross-effects occur when an indication of unbalance at one end of the rotor is actually caused by the unbalance at the opposite end. It is due to the cross-effect that unbalance indications at each end of the rotor do not represent the unbalance in the respective balancing planes.

Each indication is a result of the unbalance together with the cross-effect of the other plane. This phenomenon makes the two-plane method more complex than the single plane method. If it were not for the cross-effect, single plane balancing could have been done at each plane. Unfortunately, this does not work.

The two-plane vector method is described below. With the single-plane method, we had to make a minimum of two runs to obtain the required data. The two-plane method necessitates a minimum of three trial runs to acquire enough information for balancing correction. The procedure is briefly described as follows:

- The machine is started and the original amplitude and phase readings are recorded at each of the planes.
- A trial weight is selected and attached at the first plane. The weight and its phase angle should be recorded. The machine is started again and the amplitude and phase angle at each plane are measured and recorded.

The trial weight is then removed and installed on the other balancing plane. The machine is started once more and the amplitude and phase angle are recorded.

By using the results of the above three trial runs, it is possible to solve a 2×2 complex matrix of influence coefficients. Polar notation is useful in expressing complex numbers. The matrix equation is: $A \cdot x = B$, where A is the 2×2 influence coefficient matrix; x is the 2×1 correction mass vector; B is the 2×1 vibration reading vector; (A, x and B are all expressed in terms of complex numbers).

After this step, set vector $B = 0$ and solve vector x. The resulting vector x gives the correction mass and phase angle at each correction plane.

6.2.4 Influence coefficients

Once machines have been balanced and the data have been recorded, it can be used for subsequent balancing. The advantage is that the data can eliminate a trial run to evaluate the effect of a trial weight. The trial weight effect on a rotor supplies the information about the change in vibration that occur due to an increase or change of mass. Thus:

$$\text{weight constant} = \frac{\text{change in weight}}{\text{change in vibration}}$$

If the trial weight is increased from 2 to 4 g and the vibrations reduced from 0.3 to 0.6 mm, the weight constant would be:

$$\text{weight constant} = \frac{4 - 2}{0.6 - 0.3} = 6.67 \text{ g/mm}$$

Should it become necessary to balance the rotor of a machine whose weight constant is known, then all that needs to be done is to multiply the amplitude of vibration due to unbalance, and determine the correct weight that must be attached to the rotor. This is also called the *balance response coefficient*.

Each rotor has a different unbalance constant, given in units of weight per amplitude (e.g. lbs/mil, oz/mil, gm/in./s). A single plane balance procedure will produce one balance response coefficient. A dual plane procedure will have four coefficients. Multiple plane balancing will produce any number of coefficients depending on the number of balancing planes. For example, a three-plane balance procedure will produce nine balance response coefficients. This relation is caused by the fact that each balancing plane yields two response coefficients.

The next piece of information required relates to the phase. The phase data are converted to another coefficient called the *flash angle*. The flash angle of a rotor is the position of the heavy spot relative to the position of the vibration pickup. This angle is measured in the direction of shaft rotation. On a balanced rotor, the heavy spot is directly opposite the position where the correction weight was attached (Figure 6.4).

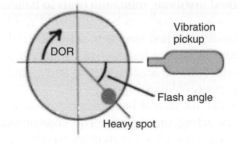

Figure 6.4
Influence coefficients

If we have successfully completed one balancing, we can follow the following procedure to determine the flash angle:

Step 1:   Record the angular position of the reference mark when the machine is running.

Step 2:   Turn the machine off and manually rotate the shaft until the reference mark is at the same location as seen while it was rotating.

Step 3:   Locate the heavy spot by finding the position directly opposite the location of the balancing weights that were applied.

Step 4:   Measure the angle between the pickup and the heavy spot, moving in the direction of rotation of the shaft.

Now that we have the flash angle of the rotor of the machine on record, we could apply it in the following manner:

- Run the machine and record the vibration and phase.
- Turn the machine off and manually rotate the shaft until the reference mark is in the same location as seen while running.
- Measure the flash angle, beginning from the actual position of the vibration pickup in the direction of shaft rotation.
- This position indicates the location of the heavy spot. A correction weight must be applied exactly opposite this location.

It is important to remember that the coefficients are valid only when the pickup is placed in the same location every time, and when the readings are taken at the same speed.

These methods are the classical first principle balancing methods. However, with the arrival of analyzers and specialized balancing software, field balancing was made much simpler because all calculations are done within the analyzer. A photocell pickup with an accelerometer is adequate for field balancing. The data is entered into the software that provides a suitable output. Dual channel analyzers can effectively do single and two-plane balancing, using influence coefficients. It is also possible to use a dual channel analyzer to conduct two-plane balancing without influence coefficients. This allows the elimination of trial runs.

6.2.5   Principle of one-step balancing using dual channel analyzers

When the number of balancing planes increases, the classical approach requires a higher number of trial runs to determine the influence coefficients. For smaller machines in a general purpose application, it may not be of much consequence. However, when it comes to critical machinery, it could take hours to conduct a single trial. The demand for multi-plane balancing is also a requirement for critical machines. Hence there is a need for a quick method involving minimum trials to balance machines.

For machines whose rotors can be considered rigid, there is a technique through which balancing is possible in just one-step, with no trial runs. This method requires a dual channel analyzer, an impact hammer (a hammer that can measure the magnitude of the impact force), one or more accelerometers and a photo tachometer (or any other means to generate a once-per-revolution signal).

We have seen that the aim of balancing is to determine the magnitude and orientation of unbalance forces acting on a plane. The conventional method is to do trial runs with known weights and then attach correction weights at the proper locations to compensate for unbalance. The influence coefficients can help reduce the number of trial runs for fine-tuning and subsequent balancing.

In one-step balancing, the trial weight runs are replaced by controlled loading of the machine structure at the transducer locations, typically by hitting the machine with an impact hammer. The analyzer can measure and compare the input load (force) and the output response (vibration) simultaneously using cross-channel analysis functions. This yields magnitude and phase relationships between force and vibration over a range of frequencies, including the synchronous components of the machine.

This information allows an accurate estimation of the imbalance loads that can be expected from vibration measurements during operation. The impact should be carried out when the machine is stopped so the correction weights can be applied. The machine only has to be stopped once. In the case of a single correction plane, the measurements and calculations are simple and straightforward. If more correction planes are used, the procedure is still straightforward, but the calculations are more complex. However, the procedure is simpler than the influence coefficient method of balancing for multi-plane applications.

6.2.6 Use of balancing machines vs field balancing

The techniques discussed earlier are associated with field balancing. During field balancing the rotor that must be balanced is mounted on its own bearings and driven at its normal operating speed. Rotors are also balanced using balancing machines. In this case, the rotor is driven by an electrical drive connected by a coupling or belt.

Balancing a rotor with balancing machines is performed during the manufacturing process, after the rotor is fully fabricated and prior to final assembly into its housing. It compensates for manufacturing errors to ensure a smooth operation after startup. Rotors are also balanced with balancing machines after repairs.

During field balancing, the rotor does not have to be removed from the housing. Removal can cause a considerable amount of downtime. Field balancing usually results in lower vibration levels because the balancing process is executed at the final operating speed, with the machine's own bearings and drive system. Some on-site factors, such as aerodynamics, misalignment and structural effects, are also accommodated.

The drawback is that it is less convenient for the person doing the balancing because the balancing machine must be transported to the site. Furthermore, there are risks involved, such as loose balancing weights being thrown from a high-speed rotor.

6.2.7 Balancing machines

There are four basic types of balancing machines:

1. Static balancing stands
2. Hard bearing machines
3. Soft bearing machines
4. High-speed machines.

Static balancing stands do not require spinning, but can correct static or single-plane unbalance only. They are sensitive enough for grinding wheels. The advantage is their low cost and safe operation.

Hard bearing balancing machines have stiff work supports, lower sensitivity and more sophisticated electronics. They require a massive, stiff foundation where they permanently stand and are calibrated in that position. Background vibration from adjacent machines can influence their results. They are used mostly in production operations where a fast cycle time is required (Figure 6.5).

Figure 6.5
Hard bearing balancing machine (IRD balancing machines)

Vertical balancers (Figure 6.6) are perfect for applications that have high-reliability requirements, high-accuracy requirements, or for rotors without journals such as clutches, flywheels, fans and blowers.

Figure 6.6
Vertical balancing machine with a drill head attachment (Schmitt balancing machines)

Soft bearing balancing machines have flexible work supports, high sensitivity and simple electronics. They can be placed anywhere and can be moved without disturbing their calibration settings. Their flexible work supports provide natural isolation; hence nearby shop floor activity can continue while the machine still achieves fine balance levels. A belt-driven soft-bearing balancing machine can always achieve finer balance results than a hard bearing machine. Every repair facility should have a soft bearing balancing machine and perhaps a static balancing stand (Figure 6.7).

High-speed balancing of bladed rotors is usually performed in a vacuum chamber to avoid high turbulence power loss. A vacuum chamber with integrated burst protection makes it possible to high-speed balance and spin-test small- to medium-sized turbo-rotors on the shop floor (Figure 6.8).

High-speed balancing involves the risk of damaging or even destroying the rotor during balancing. There are some innovative designs with unique features available to absorb the energy released when a rotor burst occurs. Safety 'crush zones' are employed and easily restored in the event of a significant burst. Balancing machines can typically balance rotors up to 8 tons, up 1.7 m in diameter and at speeds of up to 60 000 rpm.

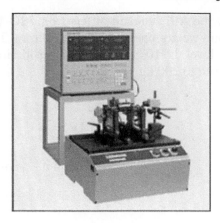

Figure 6.7
Soft bearing balancing machine (Shimadzu balancing machines)

Figure 6.8
High-speed balancing machines (Schenk Trebel)

6.2.8   Balancing limits

ISO 1940 is probably the most widely followed balancing standard. This standard defines the balancing quality grade (G) as:

$$G = e \cdot \omega$$

where e = eccentricity in mm; ω = angular velocity of rotor in radians/second.

The standard specifies nine balancing grades that are applicable to various applications. The grades are stated as G 0.4, G 1, G 2.5, G 6.3, G 16, G 40, G 100, G 250 and G 630.

The G 0.4 is the most strict grade and applies to rotors of gyroscopes, spindles and armatures of precision grinders, among others. The most lenient grade is G 630 and it is applicable to crankshaft of large and rigidly mounted diesel engines. The most commonly used grades for normal applications are G 2.5 and G 6.3. The former is applicable to most turbomachinery rotors and the latter is applicable to fans, pumps, motors and general machinery.

The American Petroleum Institute (API) specifies the balancing grade in a different way that is stricter than the ISO standards. Industries pursuing higher reliability standards from their machines are now demanding the use of API norms.

Let us see how the API standard differs from ISO 1940. Assume we need to balance a rotor of 400 kg and we are operating at 10 000 rpm. The cause of unbalance is situated at 50 mm and the level is 10 g (0.01 kg). Considering a half plane, the weight to be considered is halved to 200 kg.

Eccentricity:

$$e = \frac{0.01 \times 50}{200}$$
$$= 0.0025\,\text{mm (2.5 microns)}$$

Angular velocity:

$$\omega = \frac{2 \cdot \pi \cdot \text{N}}{60}$$
$$= 2 \times 3.142 \times \frac{10\,000}{60}$$
$$= 1047\,\text{rad/s}$$

The ISO balancing quality grade:

$$\text{G} = e \cdot \omega$$
$$= 1047 \times 0.0025$$
$$= 2.6$$

ISO 1940 specifies G 2.5 for turbomachines and thus our example is fairly in accordance with the standard. The API refers to balancing quality in terms of the residual unbalance in the rotor. It specifies the maximum limit based on the following formula:

$$\text{U-max} = 6350 \times \frac{W}{N}$$

where U-max = the maximum residual unbalance allowed per plane (g-mm); W = the weight of the rotor in kg; N = speed in rpm.
Thus:

$$\text{U-max} = 6350 \times \frac{200}{10\,000}$$
$$= 127\,\text{g-mm}$$

To calculate the eccentricity we need to divide the residual unbalance by the weight of the rotor applicable per plane, which is 200 kg in this case.
API eccentricity is:

$$e = \frac{\text{U-max}}{\text{W}}$$
$$= \frac{127}{200}\,\text{gm-mm/kg (microns)}$$
$$= 0.63\,\text{microns (0.000 63 mm)}$$

Hence, the API requirement for our rotor in terms of ISO standard will be:

$$G = e \cdot \omega$$
$$= 0.000\,63 \times 1047\,\text{rad/s}$$
$$= 0.66$$

This value is equivalent to ISO G 0.66. We thus note that the API is almost four times stricter than ISO! These balancing standards are proficient for specifying rotors that can be considered for low-speed balancing. However, it is advisable to balance turbomachine rotors at high speeds. This helps to ensure that the rotor response (maximum vibration) due to unbalance is within limits at its operating speeds and also while passing through critical speeds.

6.3   Alignment

Along with unbalance, misalignment is another major cause of unwanted vibrations. Misalignment can be internal or external with respect to a machine. Internal alignment refers to the co-axiality of bearings with respect to each other. This is accomplished by the bearing housings, casings, supports and other components, and depends on the construction of the equipment. The primary requirement is that the shaft rotation should be as concentric as possible. A good internal alignment eliminates stresses and thus ensures a smooth, vibration-free operation.

Rotating machines that are installed in pairs or trains are either driven machines or drivers. Just like the importance of internal alignment, shaft axes of two machines in a coupled condition must also be as collinear as possible during their normal operation. Often, the colinearity of shaft axes is not ensured, whether with couplings or otherwise. This is especially the case when the shafts have considerable sags. Under these conditions, colinearity of shaft axes cannot produce good alignment. This will result in vibrations.

It is also important that the alignment is achieved for the operating and not the stationary conditions. The alignment of the shaft ends can always be influenced during normal operation due to following reasons:

- Thermal expansion of supports
- Stresses from pipes
- Deformation of structures
- Modifications in the sag of the rotor with a rise in temperature.

The movement of the shaft ends due to these causes can be calculated in some instances. These calculations should be incorporated. For instance, many machines have to be kept misaligned in cold conditions so they are aligned during normal operating conditions when they become warmer.

When the causes of misalignment are beyond control, it may not be possible to obtain perfect alignment of machines. It is then recommended that they operate at the nearest possible required condition, like the required pressure and temperature in pipes and casings of pumps, compressors, turbines, etc.

6.3.1   Consequences of misalignment

Misalignment of a machine causes frictional and bending forces inside the coupling. This leads to abnormal stresses on the anti-friction bearings, as well as wear and heating of the coupling. The shaft can also fail due to fatigue caused by higher than the design cyclic stress. Cyclic stresses on bearings create axial and radial vibrations. In addition to the overloads, they cause vibrations that are transmitted to other parts of the equipment. Pumps and turbines that are equipped with special seals are particularly susceptible to failure due to excessive vibrations.

Vibrations can be transmitted to standby machines and can damage their bearings (false brineling) and also their seals. Thus, when they are required even the spare machines are not available. To avoid all the problems associated with misalignment, it becomes essential to conduct this maintenance procedure in absolute earnest. However, it should be noted that all machines do not require the same level of accuracy in alignment. It depends on the type, speed and reliability expected from a machine that determines the effort required aligning it.

6.3.2 Factors that influence alignment procedure

Influence of eccentricity (runout)

The alignment of shafts is conducted by using reference points situated on the shaft ends, coupling hubs or any other part attached to the shaft. Imperfections due to machining can cause shaft ends to be cam-shaped or eccentric. To check this, a dial gage is fixed to shaft end A with the pointer on the other end B. When end A with the dial base is rotated, the reading on the dial pointer indicates the eccentricity of end B.

Such eccentricities can result in erroneous readings and thus lead to incorrect alignment. The remedy is to couple both ends and rotate them together to obtain the misalignment of the ends. This procedure omits the error of runout and eccentricity of shaft ends or coupling hubs.

Influence of baseplate of machines (soft foot)

It is essential that the foot supports of the machines are mounted flush with the baseplate to prevent any deformation of the body or the breakage of a leg. Likewise, if all four supports are not on the same plane, the position of the axes will depend on the order in which the hold-down bolts are tightened. The contact between the baseplate and the feet can be checked with a set of shims or with feeler gages.

During a new installation, or after a modification or a revision of the baseplate, it is essential to use accurate straight edges and levels to make sure that all feet of the machine are on the same plane. The support planes of machines that will be coupled must be as parallel as possible. The accepted tolerance level for these planes is usually 0.1 mm. In some heavy machines, it may not be possible to detect a gap under the feet, even when the feet are not in the same plane.

A simple test for soft-foot is by setting up the dial gages with some residual eccentricity (as described in a previous section). In this case, a shim must be placed under one front foot and the reading noted. It is then removed and placed under the next front foot. The reading should be the same. The same procedure must be repeated for the rear feet.

Another method is to set up dial gages as before, but the dial pointer should be placed in the vertical direction. The machine is then tightly bolted to the base. Now each bolt is loosened one at a time and difference is recorded from the dial gage. Any dial reading, which is prominently higher than the others, is an indicator of a soft foot at that location. A soft foot can be fixed by placing a shim of appropriate thickness (more than the gap) under it.

Influence of axial position of machines

The axial position of shaft ends is referred to as the *distance between shaft ends* (DBSE). Normally, most couplings allow a large tolerance in the axial position and therefore it is not an important parameter to check. However, for couplings like disk couplings, an error in the axial position places the discs under stress and decreases their life. It may also generate axial thrusts, which ultimately add extra load to the machine's thrust bearings.

It is therefore necessary to take this aspect into consideration, especially when machines operate at high temperatures.

If X is the recommended DBSE for the coupling, X + 0.5 mm is used for machines conveying a product in the temperature range of 100–200 °C. It should be X + 1 mm when the conveyed product temperature is between 200 and 250 °C.

When installed on motors, it should be at zero axial deflection with the motor in the center of its float. Motor rotor float extremities and the float center should be marked on the shaft of the uncoupled motor. After marking, join them up and apply thrust in both directions. It must be confirmed that neither the inboard nor the outboard bearing stops are rubbed.

Influence of bracket

In most cases, particularly with spacer couplings, a sag check should be done on the indicator bracket to be used for the alignment. The DBSE in these couplings may be long, and when alignment brackets are clamped to one hub and extended to the other hub, there is a tendency for them to sag. This sag can alter the dial gage readings, leading to misinterpretation and errors. For bracket lengths larger than 25–30 cm, it is essential to provide additional stiffness to minimize sag.

It is therefore necessary to perform a sag check of the bracket. A sag check is essential only for aligning horizontal machines, because the sag is caused by gravity due to the weight of the bracket. In vertical machines, the bracket sag is uniform for the entire rotation of the bracket, and therefore it does not need to be checked.

To perform a rim sag check, mount the bracket on a stiff pipe, zero the indicator at mid-range in the top position and roll the pipe from top to bottom on saw horse supports. Now note the bottom reading. The total sag is twice the bracket sag. It is possible that the pipe used for a sag check can sag in itself. To restrict this problem, the following span lengths should be followed for different sizes of schedule 40 pipes:

Pipe (in.)	Span between Supports	
2	0.76 m	2 ft 6 in.
3	0.91 m	3 ft
4	1.06 m	3 ft 6 in.
6	1.32 m	4 ft 4 in.

For a face-mounted bracket, do the same type of check between lathe centers, or mount inside a pipe capped at one end. The latter method will usually require a sign reversal, but the values will be correct.

The dial reading on top is set to zero. Now the pipe is rotated 180° and the dial readings are rotated. Note the new reading. If the reading is −0.4 mm, the actual sag is 0.2 mm for the bracket and the readings taken on the machines should be fixed. In a later section, we shall see how to account for sag.

6.3.3   Alignment techniques

There are many methods to align a machine. The appropriate method is selected based on the type of machine, rotational speed, the machine's importance in production, the maintenance policy and alignment tolerances.

It may be possible to use merely a straight edge to align machines rotating at less than 1500 rpm, as well as for machines in the lower horsepower range, which are not fragile in their construction. Considering all aspects, the accuracy can be in the range of 0.3–0.8 mm.

Machines operating at speeds of 3000 rpm and higher, which are in the medium power range of 20 kW–1 MW and installed with fragile components like mechanical seals and expansion bellows, should be aligned within 0.1 mm. This requirement necessitates the use of comparators like dial gages, and methods with minimum residual errors. A majority of machines fall in this category, and hence the available methods are explained below.

Alignment conventions using a dial indicator

The dial gage is the most common comparator used during alignment. The dial gage functions on the rack-and-pinion principle. The conventions that are followed are shown in Figure 6.9.

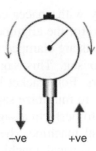

Figure 6.9
Dial indicator

When the spring is compressed, the dial pointer is pressed inward and the clock needle moves clockwise, indicating a positive reading. When the pointer moves outwards, the clock needle moves counter clockwise, indicating a negative reading.

The dial gage is normally clamped to a bracket, and the dial pointer is placed on the shaft end or coupling hub. The dial pointer is pressed to somewhere in its mid-range. This is necessary to enable positive and negative readings without encountering the end-of-travel of the pointer. At this position of pointer, the dial clock scale can be turned to match zero with the clock needle.

It is recommended to jog the pointer from the top to ensure that it is not stuck, and that repeatable readings are obtained. Another convention for alignment readings in the horizontal plane is shown in Figure 6.10.

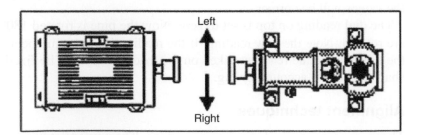

Figure 6.10
Alignment readings in the horizontal plane

Thus, the convention maintains left and right when standing behind the driver, facing the driver. Left and right readings on the dial gage are recorded accordingly. If this convention is not implemented, it should be mentioned on the graph where these readings were recorded.

Shaft setup for alignment

The connection to the shaft must be simple and rigid. The clamp shown in Figure 6.11 is a good example. Magnetic clamps must be avoided, because their attachments are not reliable.

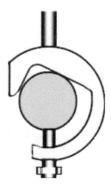

Figure 6.11
Shaft setup for alignment

There are many types of alignment brackets available in the market, and a typical one is shown in Figure 6.12. The guiding principle for the selection of brackets is that they should be rigid with minimal sag.

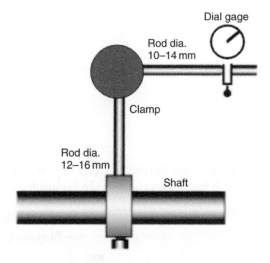

Figure 6.12
Alignment brackets

Types of misalignment

Misalignment in machines is due to angularity and offset, but in almost all cases the misalignment of machines is a combination of both.

Angularity is the difference between the values on the comparator for a half revolution. For a given angular misalignment, angularity depends on the diameter described by the dial gage. It can be seen that when d1 increases to d2, p1 increases with the same ratio to p2. This value must be fixed when a certain tolerance is given (Figure 6.13).

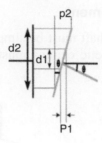

Figure 6.13
Angularity (parallelism)

Angle of misalignment:

$$\theta = \tan^{-1}\left(\frac{p1}{d1}\right) = \tan^{-1}\left(\frac{p2}{d2}\right)$$

where p1, p2 = dial gage reading when rotated by 180°; d1, d2 = diameters described by the dial gage.

When the dial gage measures concentricity, the offset becomes the radius of rotation for the dial gage, as indicated in Figure 6.14. The dial gage readings would indicate the diameter, and hence should be reduced by half to obtain the true offset reading.

$$\text{offset} = \frac{\text{dial gage reading}}{2}$$

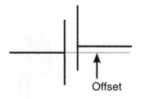

Figure 6.14
Radial misalignment (concentricity)

However, as mentioned before, in practice misalignment of machines is due to a combination of both factors, as depicted in Figure 6.15.

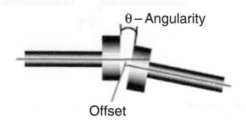

Figure 6.15
Misalignment of shafts with angularity and offset

Two dial method of alignment

The necessary steps to align a machine are:

1. The first step is to loosen the coupling bolts so there is no restriction during the measurement of angularity of the existing misalignment. A feeler gage is then run through the coupling hubs to ensure that the hubs are not touching.
2. The dial gage is attached as shown in Figure 6.16. The first step is the radial test to measure the offset. This is done in the vertical and horizontal planes. To obtain the offsets in both planes, four readings will be required. The convention that should be used is either-

  - Top, bottom, left and right (left/right convention is as indicated earlier)
  - Clock positions – 12 o'clock, 3 o'clock, 6 o'clock and 9 o'clock positions.

  The dial gage is generally placed on the *top* (12 o'clock) position, and the zero on the scale is turned to coincide with the needle. The pointer must be jogged to ensure that it is free and that the readings are repeatable.

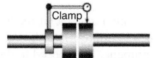

Check for Concentricity

Procedure – Rotate the dial gage
by 180 degrees and note the initial
and final readings.

Figure 6.16
Dial gage setup at top position. The difference in readings after 180° indicates offset in vertical or horizontal planes

  In this position, both shafts are turned manually through one complete revolution, and readings at every quadrant are noted. The readings recorded at the four locations are written down in the format shown below. The 'R' in Figure 6.17 indicates that these are radial readings, meant for offset corrections.

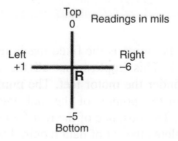

Figure 6.17
Readings in mils

3. The clamp is readjusted with the dial gage pointer now set to measure the angularity, as shown in Figure 6.18. The pointer (as shown in the figure) is now parallel to the axes of the shafts. Just like the offset was measured in both the horizontal and vertical planes, the angularity must be measured in both planes as well. The dial gage is rotated once more through one complete revolution and stopped at every quadrant to make a note of the readings.

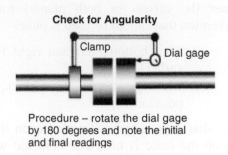

Check for Angularity

Procedure – rotate the dial gage
by 180 degrees and note the initial
and final readings

Figure 6.18
Dial gage setup at top position. The difference of readings after 180° indicates angularity in vertical or horizontal planes

4. Once these radial and facial readings are recorded, the next step is to convert these values to appropriate shim thickness that should be added or removed to fix the alignment. To proceed to the next step, additional information about the location of the front and the rear feet from the dial gage pointer is required (Figure 6.19).

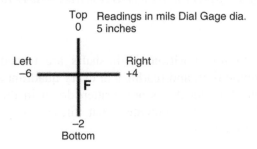

Figure 6.19
The 'F' indicates facial readings (note the diameter described by the dial gage)

In Figure 6.20, the pump is the fixed machine (FM) and the motor is the machine to be shimmed (MTBS). This implies that all the corrections will be done by adding and removing shims under the motor feet. The pump will not be disturbed from its position. The distance from the pointer of the dial gage to the front foot (FF) of the motor is designated as 'A'. The distance of the rear foot (RF) to the dial gage pointer is designated as 'B'. This completes the set of data required to perform the necessary calculations. Two sets of calculations are required. One set for the vertical plane and the other for the horizontal plane.

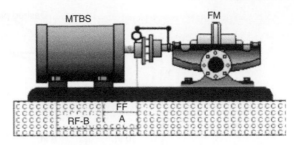

Figure 6.20
Shimming Calculation

Calculations for the vertical plane

Offset correction   For example, let us say the offset readings for the top and bottom positions are 0 and −5 mils, respectively. If the dial gage pointer is on the motor (MTBS), the negative sign indicates that the motor shaft is higher than the pump shaft. It is higher by half the final reading minus the initial readings. Thus:

$$\frac{(-5)-0}{2} = -2.5 \text{ mils}$$

Hence, shims of 2.5 mils should be removed from the front and rear feet.

Angularity correction   Let us say the angularity readings for the top and bottom readings were 0 and −2 mils, respectively. If the dial gage pointer is touching the rear face of the motor coupling hub, the negative sign indicates that the coupling has a narrower gap at the bottom than at the top. The dial scribes a circle of 5 in. The angle $\theta = \tan^{-1}(p1/d1)$. Because the angle is very small, the tan inverse function can be neglected:

$$p1 = 0-(-0.002) \text{ in.}$$

(The formula would reverse if the pointer is touching the front face of the coupling hub, which is normally the case when there is a long spacer between the couplings.)

$$d1 = 5 \text{ in.}$$
$$\therefore \theta = \frac{0.002}{5}$$
$$= 0.4 \text{ milli-radians } (0.023°)$$

This angle θ is also the angle of inclination of the motor axis with respect to the pump axis. The triangle of angularity at the coupling is similar to the triangle of motor inclination. Line AB is the existing axis inclination of the motor (Figure 6.21). It must be lifted by amount x at the FF (front foot) location and by y at the RF (rear foot) location. The x and y values are calculated as follows. x and y are approximated as arcs and the following formula can be used:

$$s = r \times q$$

where s = arc length; r = radius; θ = included angle.
Hence:

$$x = 8 \times 0.4 \ = 3.2 \text{ mils (add shims)}$$
$$y = 18 \times 0.4 = 7.2 \text{ mils (add shims)}$$

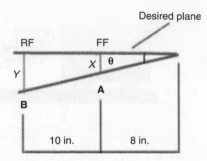

Figure 6.21
Calculating X and Y values

The final results should include corrections for both the offset and the angular corrections.

At point A:
Offset results – remove shims of 2.5 mils
Angularity results – add shims of 3.2 mils

Thus, insert shims of 0.7 mils under the front foot of the motor.

At point B:
Offset results – remove shims of 2.5 mils
Angularity results – add shims of 7.2 mils

Thus, insert shims of 4.7 mils under the rear feet of the motor.

Calculations for the horizontal plane

To revise the procedure, we repeat the calculations for the horizontal plane. The convention is: from behind the motor, left is the initial reading and right is the final reading.

Offset calculations:
Left reading: +1 mils
Right reading: −6 mils
Because the dial pointer is on the motor shaft, a negative right reading indicates that the motor shaft axis is to the left of the pump shaft axis.

$$\text{Offset} = \frac{-6 - (+1)}{2} = -3.5 \text{ mils}$$

Move points A and B of the motor to the right by 3.5 mils.

Angular calculations:
As the dial pointer touches the rear face of the motor coupling hub, the shaft axis resembles what is shown in Figure 6.22.
In this case:

$$p1 = +4 - (-6) = +10$$

$$d1 = 5 \text{ in.}$$

Thus:

$$\theta = \frac{0.01}{5}$$

$$= 2 \text{ milli-radians } (0.114°)$$

Hence: $x = 2 \times 8 = 16$ mils – move to the left; $y = 2 \times 18 = 36$ mils – move to the left.

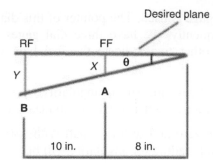

Figure 6.22

At point A:
Offset results – move 3.5 mils to the right
Angularity results – move 16 mils to the left

Thus, move to the left by 12.5 mils.

At point B:
Offset results – move 3.5 mils to the right
Angularity results – move 36 mils to the left

Thus, move to the left by 32.5 mils.

The vertical shim corrections should always be done prior to the horizontal shifts. Once the vertical shims are adjusted, the bolts should be tightened and a quick test of the vertical plane reading should be made to confirm the accuracy. If the accuracy is satisfactory, the bolts can be loosened and the horizontal alignment should be done with jack bolts (if provided). The limitations of this method are:

- Calculations are necessary, which may be difficult to do in the field.
- It is beneficial to be able to visualize the shaft orientation from the dial gage readings but this requires practice. Inexperienced technicians can find this confusing.
- It is prone to errors in calculations, bracket sag and the dial gage readings.
- If the shaft of one or both the machines have substantial axial floats, the angular readings can be erroneous.

Three dial method of alignment

In the section dealing with the two-dial method, we used only one dial gage but with two measurements, one with the pointer in the radial direction and the other parallel to the shaft axes. There are brackets available which can accommodate two dial gages at a time that can measure the offset and the angularity of misalignment. This reduces the time required to collect readings. In the limitations of the two-dial method, we came across a special case of alignment of machines that have a high axial float. These can include:

- Floating shaft engines (axial float is 10 mm)
- Machines with Kingsbury-type thrust bearings (axial float is 0.2–0.3 mm)
- Machines with worn out antifriction bearings or tapered roller bearings (0.05–0.1 mm).

If the shafts of these machines move axially while manually rotating them to obtain facial readings, erroneous readings might be obtained. To eliminate this error an additional dial gage is used. The pointer of this dial is placed at 180° from the other dial pointer. Consequently, we have three dial gages (Figure 6.23). One shows the radial offset and the other two indicate the facial or angular readings. We define them as follows:

> Fm – dial gage measuring angularity
> Fr – dial gage used as reference to measure angularity (pointer 180° apart).

Set the gages to zero and turn the shaft with 180°. Record both the dial gage readings. The correct facial reading for angularity will be:

$$F = \frac{Fm - Fr}{2}$$

Example:

Fm and Fr are set to zero. The shafts are rotated with 180° and the new readings are noted. If Fm = −4 and Fr = −20, the angularity will be:

$$F = \frac{-4 - (-20)}{2} = +8$$

Once the correct angularity is obtained, the rest of the procedure is identical to the two-dial gage method.

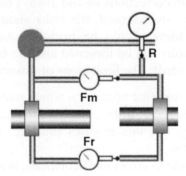

Figure 6.23
A setup showing three-dial alignment setup

Reverse dial method of alignment

The setup for the reverse dial method is shown in Figure 6.24. The reverse dial method is generally employed on dual articulation couplings. These are couplings that have spacers between them. This method provides many advantages, listed below:

- Accuracy is not affected by axial movement of shafts.
- When both the shafts are rotated together, runouts on coupling hubs are not measured.
- Geometric accuracy is better than the two-dial method.
- Couplings do not have to be opened to measure the misalignment.
- Sag and thermal rise corrections can be incorporated.

Consider the assembly in Figure 6.25 to obtain shim corrections with the reverse dial method.

The following conventions are used:

- Dial B with pointer on the fixed machine (not shown) is in plane B
- Dial A with pointer on the motor (MTBS) is in plane A

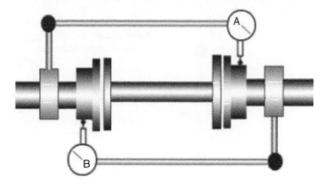

Figure 6.24
Reverse dial method

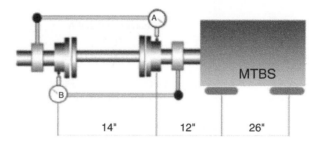

Figure 6.25
An assembly employing the reverse-dial method

- Motor foot closer to dial A is the inboard foot and named as IB foot
- The outboard foot of the motor is called the OB foot
- Distance between plane A and B = 14 in.
- Distance between plane A and IB = 12 in.
- Distance between IB and OB = 26 in.

Upon checking for sag the following observations were made. The dial with the clamp is set as shown in Figure 6.26 in the top position. The pipe with the clamp is then rotated through 180°. Due to the sag of the clamp, the dial pointer moves outwards, indicating a negative value of −6 mils. Dial gages A and B are then connected to the machine as shown in the figure. Dial A has its pointer on the motor coupling hub and is in the *top* position. The pointer of dial B is on the fixed machine and is in *bottom* position. The shafts are then turned through one complete revolution and readings of the dial gages are recorded at every quadrant. These readings are shown in Figure 6.27.

The readings have been influenced by sag. The sag check indicated a dial reading of −6 mils. This was obtained after rotating the dial from the top to the bottom (TB) positions. Dial gage A has moved from the TB position too. This implies that it is showing 6 mils more than the value due to misalignment. This value should therefore be subtracted to obtain the correct reading.

The dial A readings are: top: 0; bottom: +20. After sag correction the readings should be: top: 0; bottom: 14. The horizontal readings, as mentioned before, are not affected by sag.

Dial A is on the motor (MTBS) shaft. The bottom reading is positive, indicating that at the vertical plane of the dial A pointer (say plane A), it is in a higher position.

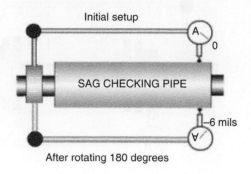

Figure 6.26
Dial with clamp rotated 180°

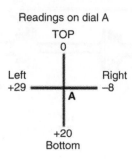

Figure 6.27
Readings showing sag on dial A

The TB readings of dial B (Figure 6.28) must also be corrected for sag. In this case, the dial begins its journey at the bottom and finishes on top. For this reason, technicians normally keep the dial reading on zero at the bottom position. The procedure of sag correction can get confusing here. The author recommends that the above procedure of subtracting the sag check reading (6 mils in this case) be repeated for the bottom reading.

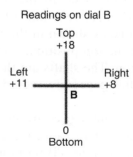

Figure 6.28
Readings showing sag on dial B

The sag corrected readings of dial B are: top: +18; bottom: −6. Now add +6 mils to both the top and bottom readings: top: +24; bottom: 0.

The pointer of dial B is on the fixed machine (Figure 6.29) and while moving from the top to the bottom the pointer gets pressed inwards to yield a positive reading. This implies that in the vertical plane of the dial B pointer (say plane B), it is also higher.

Hence, the dial readings that must be used for calculations are:

dial A: top: 0   bottom: 14 position in plane: higher
dial B: top: 24 bottom: 0   position in plane: higher

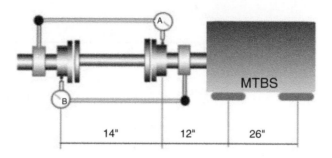

Figure 6.29

The indicated dial gage readings are double the misalignment values, hence:
   radial offset at plane A is 7 mils
   radial offset at plane B is 12 mils
The distance between planes A and B is indicated as 14 in. Consequently, the angular misalignment is obtained with (Figure 6.30):

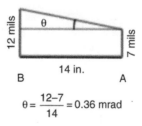

$$\theta = \frac{12-7}{14} = 0.36 \text{ mrad}$$

Figure 6.30

Once the angle is known, the line is extended to OB as shown in Figure 6.31. The distances at IB and OB can be determined from the graph, and these values indicate the shim thickness that must be removed or inserted.

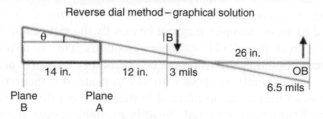

Figure 6.31
Graph showing line extended to OB

The horizontal movements are calculated in a similar way.

The dial A readings are: left: +29   right: −8
This can be reduced to: left: +37   right: 0
The dial B readings are: left: +11   right: +8
This can be reduced to: left: 0   right: −3
Radial offset (Figure 6.32) in horizontal plane A (dial A) = 18.5
    pointer position in plane: left
Radial offset in horizontal plane B (dial B) = −1.5
    pointer position in plane: right

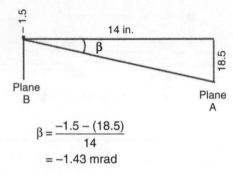

$$\beta = \frac{-1.5 - (18.5)}{14}$$

$$= -1.43 \text{ mrad}$$

Figure 6.32
The angle of misalignment in horizontal plane

Thus, the IB foot has to be moved to the right by 35 mils (Figure 6.33) and the OB foot must also move to the right by 74 mils.

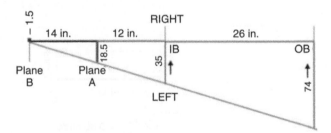

Figure 6.33
Reverse dial method – graphical solution – horizontal plane

Note – compensation for thermal growth

Shaft-coupled machines like steam turbines, pumps handling hot liquids and high-speed gearboxes operate at temperatures high enough to cause an expansion of the machine housing. This phenomenon is known as *thermal growth*. The thermal growth of machines coupled to these examples may not be exactly the same, which causes relative displacement of shafts from their 'cold' stationary positions. This in turn results in deterioration of the alignment condition, unless proper compensatory measures are employed.

It is consequently necessary to compensate for thermal growth. The compensation causes machines to be misaligned in their cold condition, but under operating conditions proper alignment is achieved. Readily available target specifications for cold alignment are generally obtainable from machine manufacturers. However, when the manufacturer does not provide thermal growth specifications, the following guidelines can be used.

The machine pedestal can be assumed to be experiencing a unidirectional thermal expansion. The formula for linear expansion is used to calculate this increase in length.

$$dL = (L \times \alpha)\, dT$$

where dL = thermal expansion; L = height centerline to base of machine; α = coefficient of thermal expansion of material (0.000 01 18 for cast iron in SI units and 0.000 00 59 in FPS units); dT = change in temp from ambient.

Example:
Consider a pump with a liquid at 120 °C. The base-to-center height is 500 mm (19.7 in.). The ambient temperature is 20 °C.

$$dL = (L \times \alpha)\, dT$$

$$dL = 500\,\text{mm} \times (0.000\,01\,18) \times (120 - 20)\ [19.7 \times 0.000\,00\,59 \times 212]$$
$$= 0.59\,\text{mm}\quad (24.6\ \text{mils})$$

There is also a rule of thumb that can be used as a quick method to determine thermal expansion. The rule states that:

- There is an expansion of 1 mm for 1 m of length for a 100 °C rise in temperature
  OR
  There is an expansion of 1 mil for 1 in. of length for a 100 °C rise intemperature
- Increase the total by 20%.

In the above example, the rise in temperature was 100 °C. Following the rule of thumb, a 0.5 m length will expand with 0.5 mm. Increase it with 20%, and the answer is 0.6 mm. In inches, let us approximate the length to be 20 in., so for 100 °C raise it will be 20 mils. Now increase the result with 20%, and the answer is 24 mils. This is very close to the answer above!

Once the thermal growth is known, it can be compensated for with shims. If the alignment results indicate that the feet have to be raised by adding shims of 0.85 mm and we see that the thermal growth will be 0.6 mm, then shims of only 0.25 mm should be added. The rest of the required rise will occur through expansion. If the alignment results indicate that shims of 0.4 mm must be added in the above case, then shims of 0.2 mm should be removed.

Laser alignment

Alignment with comparators such as dial gages provides a fair degree of precision, but these methods demand skill, training and experience. Consequently, these methods are prone to errors and can take a considerable amount of time. The method of alignment using lasers (Figure 6.34) overcomes the disadvantages listed above and it is gradually becoming the preferred method of alignment for most machines. Due to recent advances in this technique, the alignment, data collection and calculations have become fast and accurate.

Some laser systems need less than a quarter turn of the shaft to produce very good shim correction data. They have built-in alignment tolerances, and hence there is no need for an expert to judge on the quality of the residual misalignment.

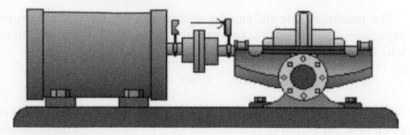

Figure 6.34
Laser alignment

In addition to the obvious advantage with critical machines, laser alignment is ideal for aligning cooling tower gearboxes with their motors. The distance between these is typically up to 2–5 m. The dial gage method for these applications is very cumbersome. Laser beams can travel over long distances, and hence alignment of machines so far apart can be done very accurately with relative ease.

The laser alignment system comprises an analyzer (Figure 6.35) and two laser heads. The laser heads are attached to the two shafts. The laser heads must face each other, and each head has a laser emitter and receiver. When the shafts are turned, the receivers trace the movement of the laser beams. These values are communicated to the analyzer. Machinery data and the required distances are initially entered into the analyzer. The data from the laser heads and the given machinery data are used to accurately determine the shim corrections for the machine. Laser beams do not bend over great distances and for this reason the sag effect is entirely eliminated.

Figure 6.35
Laser alignment system comprising of laser head, reflector and analyzer (Prueftechnik – Optalign Plus system)

Some entry-level laser alignment systems only have one laser emitter head and a reflecting prism on the other. The analyzer in this case is provided with only basic features. These systems are ideal for general purpose machines. They eliminate the dial gages and provide an alignment calculator. The methodology with these systems is the same. Once the laser head and the reflector are installed, the shafts must be rotated. At every quarter revolution, the analyzer must be activated to acquire the reading. After this, the analyzer provides the alignment correction information.

Some advanced systems include additional features that make alignment of machines an easy task. Some of these features are:

- Complex trains comprising of as many as five machines can be handled.
- RF communications eliminate cables between the laser heads and the analyzer.
- In-built inclinometers can instantly find the shaft position.
- Errors due to vibrations from other machines can also be eliminated through averaging.
- Uncoupled and non-rotating machines can also be aligned.
- Less than a quarter rotation may be sufficient to obtain misalignment data.
- It is possible to do *live* horizontal alignment. This means that there is no need to take a reading and transfer it to the analyzer for calculation. The instant communication of the heads and analyzer accomplishes this automatically.
- One or two soft foot conditions can be identified.
- Thermal growth of machine pedestals can be calculated and incorporated in calculations.
- Templates of known machinery trains are programed into the analyzer, and these can be selected for better understanding and reporting.
- Once a machine is aligned, its history and data can be stored.
- They provide built-in misalignment tolerances.

6.3.4   Alignment tolerances

In practice, it is almost impossible to obtain a zero offset and zero angularity, and thus machines have to be left with a certain residual misalignment. This residual misalignment has little or no detrimental effect on the operation of machines. The following table (source: *CSI UltraSpec – Easy Align Manual*) provides us with residual misalignment values that are safe. The values are based on the machine's operating speed.

	Excellent		Acceptable	
	Offset (mils)	**Angle (mils/in.)**	**Offset (mils)**	**Angle (mils/in.)**
Speed (rpm)				
<500	5.0	1.5	6	2.0
500–1250	4.0	1.0	5	1.5
1250–2000	3.0	0.5	4	1.0
2000–3500	2.0	0.3	3	0.5
3500–7000	1.0	0.25	2	0.3
>7000	0.5	0.2	1	0.25

These values are assumed to be pure offset or pure angle. In practice, a combination of the two is more common and tolerances should account for this combination. For example, a machine is running at 3000 rpm and the residual misalignment data is:

offset: 2.6 mils   angularity: 0.25 mil/in. (mrad)

In pure terms, these values would be acceptable. Nonetheless, let us see if the combination of the two is acceptable. To achieve this, a XY graph is made as shown in Figure 6.36.

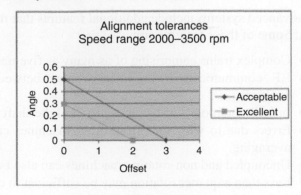

Figure 6.36
Alignment tolerances

If an offset of 2.6 with an angularity of 0.25 mils/in. is plotted, it could be beyond the acceptable range.

6.4　Resonance vibration control with dynamic absorbers

In an earlier section, we discussed the phenomenon of resonance in detail. When a sinusoidal force acts on an undamped mass-spring system and the forcing frequency is equal to the natural frequency of the mass, the response (amplitude of vibration) is infinite. This is called resonance. The problem associated with this phenomenon is that the response of the system or the amplitude of vibration increases to a very large magnitude. This can cause damage to the mechanical system.

Numerous analytical studies and tests are done during the design stages of mechanical systems in order to avoid this phenomenon during normal operation. Resonance cannot be eliminated, but it can be managed.

All mechanical systems will have a natural frequency. As a consequence they will certainly encounter resonance. If this natural frequency alters due to various factors or for some reason it happens to coincide with an operational frequency, resonance will occur. In some cases, there may be a simple solution to attenuate the system, but in some cases, expensive modifications to machines and/or its components may be essential.

For such cases, a simple and inexpensive gadget called a dynamic absorber is used. It operates on the principle explained below.

Consider the spring-mass system. Now we attach another spring-mass system in series to the first one, as shown in Figure 6.37.

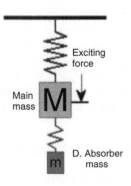

Figure 6.37
Resonance vibration control

After the absorbing mass-spring system is attached to the main mass, the resonance of the absorber is tuned to match that of the main mass. When this is achieved, interestingly, the motion of the main mass is reduced to zero at its previous resonance frequency.

Thus, the energy of the main mass is apparently 'absorbed' by the tuned dynamic absorber (Figure 6.38). It is interesting to note that the motion of the absorber is finite at this resonance frequency, even though there is *no* damping in either of the oscillators.

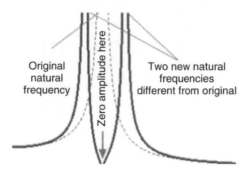

Figure 6.38
Dynamic absorber

This happens because the system changes from a one-degree of freedom system to a two-degree of freedom system. The combined mass system now has two resonance frequencies. Neither of the resonant frequencies equals the original resonance frequency of the main mass when it was alone (and also the absorber).

If the system has no damping, the response of this combined two-degree of freedom system is infinite at these new natural frequencies. This may not be a problem when a machine is not normally running at its natural frequency, but a large infinite response can cause problems during startup and shutdown.

A finite amount of damping for both masses will prevent the amplitude of vibration of either mass from becoming infinite at either of the new natural frequencies. However, if damping is present in either mass-spring element, the response of the main mass will no longer be zero at the target frequency. The two-degree of freedom system has two natural frequencies, corresponding to the two natural modes of vibration of the system.

At the lower frequency mode (frequency less than the original natural frequency) the masses move in phase with each other. The original mass (*M*) and the absorber mass (*m*) will move in the same direction during each cycle. At the higher frequency mode (frequency greater than the original natural frequency) the two masses move in opposite directions, 180° out of phase with each other. In this case, when the original mass (*M*) is moving upwards, the absorber mass (*m*) would be moving downwards and *vice versa*.

6.4.1   Designing a dynamic absorber

A dynamic absorber can be designed using a simple mathematical formula. The level of the dynamic forces exciting the resonance is not usually known, and therefore some minor field adjustments are necessary.

Material	Young's Modulus E [lb/in.^{2}]	Density m [lb/in.^{3}]
Steel	29 000 000	0.282
Aluminum	10 000 000	0.099
Copper	16 000 000	0.321
Iron	18 000 000	0.260

A sketch and calculations required to design a simple dynamic absorber are shown in Figure 6.39. It is required to determine the weight *w* that must be placed at position *a* on a piece of rectangular bar of length *L* and with a cross-section of width *b* and height *h*. The first parameter to be estimated is the freestanding length. For example, a dynamic absorber for a 3000-cpm natural frequency would be in the range of 12–18 in. This is dependent on the material and the rectangular cross-section.

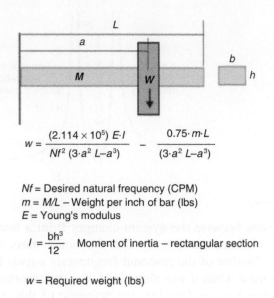

$$w = \frac{(2.114 \times 10^5)\, E \cdot I}{Nf^2\, (3 \cdot a^2\, L - a^3)} - \frac{0.75 \cdot m \cdot L}{(3 \cdot a^2\, L - a^3)}$$

Nf = Desired natural frequency (CPM)
m = *M/L* – Weight per inch of bar (lbs)
E = Young's modulus

$I = \dfrac{bh^3}{12}$ Moment of inertia – rectangular section

w = Required weight (lbs)

Figure 6.39
Design calculations for a dynamic absorber

Two dynamic absorbers of steel are tabulated below.

	3000 cpm	**3600 cpm**
b	0.75 in.	0.75 in.
h	0.5 in.	0.5 in.
L	14.0 in.	13.0 in.
a	12.0 in.	11.0 in.
M	1.48 lbs	1.375 lbs
w	0.527 lbs	0.422 lbs

	1475 cpm	**1750 cpm**
b	0.75 in.	0.75 in.
h	0.5 in.	0.5 in.
L	20.0 in.	18.0 in.
a	18.0 in.	16.0 in.
M	2.115 lbs	1.904 lbs
w	0.685 lbs	0.752 lbs

It should be noted that the above tables are case study designs. For permutations of the material properties, dimensions L, b, h and a can provide a preferred dynamic absorber. It is important that the dynamic absorber has a rectangular cross-section. Structural resonance problems are typically directional in nature. The dynamic absorber must be designed to have a natural frequency in a specified direction.

If a rod with a circular cross-section is used, the radial stiffness of the rod will be uniform in all directions, and the rod will whip in a circular or conical pattern rather than in one direction. This will reduce the amplitude of vibration in the initial direction of resonance, but it will create a resonance problem in a direction perpendicular to the original resonance. The same argument applies to a square bar.

6.4.2   Applications of a dynamic absorber

Dynamic absorbers are quite effective in controlling resonance on machines that must operate over a wide range of operating speeds. Increasing or decreasing the mass or stiffness can attenuate the unwanted vibration, but may cause further resonance problems at a new natural frequency. Dynamic absorbers are used to minimize damage to pumps, air compressors, refrigeration machines, machine tools and other machines that repeatedly pass through a structural resonance each time they are started or stopped. Another useful application for dynamic absorbers is to verify a resonance problem when other analysis techniques are not possible.

For example, it is not possible to shut down a machine and make a Bode plot or do a 'bump' test. Conversely, it may be possible to temporarily attach a dynamic absorber while the machine is operating. If the vibration amplitude reduces after tuning the absorber, it was truly a resonance problem. If the vibration problem is not due to resonance, attaching a dynamic absorber can actually create a resonance problem and effectively amplify or increase the existing vibration, making it worse than it was before the absorber was attached.

Temporary dynamic absorbers can be attached to machines using a C-clamp or by bolting them to the machine. If the vibrations are effectively reduced, these can be left in the same position until an opportunity is available to attach a permanent absorber.

7

Oil and particle analysis

7.1 Introduction

In the earlier topics, we covered the technique of vibration analysis in detail. It is undoubtedly the most important predictive maintenance technique. Oil analysis is another predictive maintenance technique that evolved through the years and is currently maturing into a reliable source of predictive machinery information.

Oil analysis is not merely a tool to analyze the condition of a lubricant. With modern diagnostic tools, it is used to monitor the condition of equipment as well. By utilising these advanced techniques, equipment reliability can increase, and unexpected failures and downtime are minimized. There are many wear mechanisms that lead to the deterioration of machine components. Though there are different *types* of wear, there are only a few primary *sources* of wear.

The mechanisms that contribute to wear of a component include misalignment, unbalance and improper use of the equipment (such as overload or accelerated heating conditions). One of the sources for wear can be the lubricant itself, e.g. in cases where the lubricant has degraded or has become contaminated.

The different types of wear that can occur are:

- Abrasive wear
- Adhesive wear
- Cavitation
- Corrosive wear
- Cutting wear
- Fatigue wear
- Sliding wear.

In an operating machine, there is a continuous attrition of its components due to the generation of wear debris. Lubricants normally carry the debris away. Identification and analysis of this wear debris can pinpoint the type of wear and also identify the source, which could be any component under distress. Oil analysis can highlight the necessity to initiate a corrective action to prevent possible breakdowns.

In this way, it becomes an important predictive maintenance technique. There have been cases where oil analysis helped to identify rotating equipment defects even before a vibration analysis could detect it. This especially applies to slow speed machines with high load levels, like diesel engines.

When implementing an oil analysis condition-monitoring program, it is important to select proper tests that will identify abnormal wear particles in the oil. The program should be customized to the type of equipment being monitored and the expected failures that are anticipated. The types of tests, location of sampling, analyses and the interpretation of oil analysis depend heavily on whether the application is a compressor, steam turbine, diesel engine, gearbox or a hydraulic system.

As mentioned, wear particles are the prime indicators of the machine's health. There are many techniques to evaluate the type and concentration of such particles. The techniques include:

- Spectrometric analysis
- Particle counting
- Direct reading ferrography
- Analytical ferrography.

In addition to wear particle analysis, it is also necessary to trend the condition of the oil itself. Good oil can reduce the generation of wear to a great extent. Thus the analysis of the oil condition is an essential part of the program. In a certain sense, it is a proactive maintenance program. Some types of oil analyses are:

- Viscosity
- Solids content
- Water content
- Total acid number
- Total base number
- Flash point.

Another aspect of an oil analysis maintenance program is the sampling of oil. In order to do a reliable oil analysis, it is essential that the collected sample is able to indicate the true condition of the machine. Any external contamination, not sourced from the machine, will yield erroneous results. Care and caution are required to ensure the prevention of external contamination.

Besides prevention of external contamination, it is also necessary that the collected samples have a good concentration of particles. This fact places a lot of emphasis on the importance of the location where the oil sample is collected. In circulating oil systems, the best location is in a live zone of the system, upstream of filters where particles from ingression and wear debris are concentrated. Usually, this will mean sampling on fluid return or drain lines.

In some cases, where the oil drains back to sumps without being directed through a line (like diesel engines or reciprocating compressors), the pressure line downstream of the pump upstream of the filter must be used. Sampling from stagnant zones such as static tanks and reservoirs is not a recommended sampling practice. Splash, slinger ring and flood-lubricated components are best sampled from drain plugs after considerable flushing, or preferably using a portable circulating off-line sampler. These factors are discussed in detail in this section. We start off with understanding the fundamentals of oil itself.

7.2   Oil fundamentals

A lubricant usually has a base fluid. The base fluid is generally of petroleum origin, combined with additive chemicals that enhance the various desirable properties for a base fluid. Base fluids are essentially obtained from two main sources. One is the refining of petroleum crude oil, and the second is the synthesis of relatively pure compounds with properties that are desirable for lubricants.

7.2.1 Mineral oils

The general principles of lubricant base oil manufacture involve a series of steps to improve certain desirable lubricant properties, such as:

- Viscosity index
- Oxidation resistance
- Heat resistance
- Low-temperature fluidity.

Starting with petroleum crude oil, the typical process for making lubricant base oil involves:

- Separation of lighter boiling materials such as gasoline, jet fuel, diesel, etc.
- Removal of impurities that include aromatics and polar compounds
- Distillation to give desired base oil viscosity grades
- Dewaxing to improve low-temperature fluidity
- Finishing to improve oxidation and heat stability.

7.2.2 Synthetic oils

Another source of lubricant base fluids is synthetic of origin. A suitable definition for a synthetic material is: 'A product prepared by chemical reaction of lower molecular weight materials to produce a fluid of higher molecular weight designed to provide certain predictable properties.'

This is in contrast with refined petroleum oils, which are composed of many compounds of varying chemical composition, depending on the refining method and the crude stock source.

The three most common types of synthetic base oils are:

1. Polyalpholefins
2. Organic esters
3. Polyglycols.

Other synthetic fluids find niche uses in very specialized applications. These include phosphate esters, silicones, silicate esters and polyphenyl ethers. Synthetic lubricants have several advantages over conventional mineral oils:

- Excellent low-temperature fluidity
- Low pour point
- High natural viscosity index
- Excellent oxidation stability
- High flash, fire and auto-ignition points
- Low volatility
- Non-corrosive and non-toxic.

Synthetic lubricants such as low-temperature lubricants and fire-resistant hydraulic fluids have been in use in the airline industry for quite some time. These are applications that justify the high cost of these lubricants. Synthetic materials can generally be used over a wider temperature range than petroleum-based fluids within the same viscosity range. Certain synthetic lubricant base stocks can be blended with petroleum oils to obtain the necessary high-temperature volatility and low-temperature viscosity characteristics when the proper petroleum-based oils are unavailable.

7.2.3 Additives

Additives are defined as materials that impart new properties to the base mineral oil. Additives enhance the existing properties of the lubricant into which they are incorporated. The amount and type of additives that are blended with the lubricant depend on the performance features desired from the lubricant. Different additives commonly used are:

Detergents (metallic dispersants)   These are used to control deposit formation throughout the system. They keep machine parts clean and hold the deposits if they are formed in a suspension. Detergents, like dispersants, are blended into lubricants to remove and neutralize harmful products. In addition, detergents form a protective layer on metal surfaces to prevent deposits of sludge and varnish. In engines, this can reduce the amount of acidic materials produced. The metallic basis for detergents includes barium, calcium, and magnesium and sodium. Typical applications for detergent additives are primarily diesel and gasoline engines.

Ashless dispersants   The purpose of this additive is to suspend or disperse harmful products in the lubricant. Thus, the additive neutralizes the effect of these products. Harmful products include contaminates such as dirt, water, fuel, process material, as well as lubrication degradation products such as sludge, varnish and oxidation products. Typical applications include diesel and gasoline engine oils, transmission fluids, power steering fluids and in some cases gear oils.

Oxidation and bearing corrosion inhibitors   Rust and corrosion are the result of the attack on the metal surfaces by oxygen and acidic products, and the process is accelerated by the presence of water and impurities. Rust and corrosion inhibitors neutralize acids and form protective films over the sliding surfaces. These inhibitors must be in the lubricant and on surfaces above the liquid level.

Antioxidants   An antioxidant limits the oxidation of oils that operate at elevated temperatures. Also known as oxidation inhibitors, they interfere with the oxidation process by chemically converting oxidation products to benign products. In addition, some oxidation inhibitors interact with the free catalytic metals (primarily copper and iron) to remove them from the oxidation process. Almost all lubricants commercially available contain anti-oxidation additives in varying degrees.

Viscosity index improvers   These additives improve the viscosity–temperature relationship of the oil. Viscosity improvers are added to a lubricant to retain satisfactory lubricating capabilities at higher temperatures. At low temperatures the viscosity characteristics of the base stock prevail while at high temperatures the viscosity improver maintains the correct viscosity.

Pour point depressants   Pour point depressants provide gravity flow properties for oils at low temperatures. They tend to inhibit the formation of wax at low temperatures. In many oil formulations, especially those containing viscosity improvers, supplemental pour depressants are not necessary since other additives also have pour point depressant properties.

Extreme pressure, anti-wear additives   These provide the necessary load-carrying capabilities for the oil and prevent scuffing of moving parts under boundary lubrication.

Foam inhibitors   Foam inhibitors control the formation of foams. Anti-foam agents are used to reduce the foaming tendencies of the lubricant. Foam inhibitors are added to a lubricant in service if a foaming problem is detected. The lubricant and equipment manufacturers should be consulted before adding a foam inhibitor as the correct quantity is of extreme importance. Excess quantities of anti-foam additives can also lead to excessive foaming.

Emulsifiers   Emulsifiers reduce the surface tension of the oil.

Demulsifiers   Demulsifiers reduce the tendency to form emulsions and assist in easy separation from water.

Mist suppressors   These reduce formation of an oil mist, which is a source of environmental pollution, and cause loss of the lubricant.

Tackiness agents   These improve adhesion and stickiness to metals.

Biocides   These control bacterial and fungal growth.

The following is a list of common elements that are found in additives and the role their properties impart on lubrication oil:

Barium (Ba)	detergent or dispersant additive
Boron (B)	extreme-pressure additive
Calcium (Ca)	detergent or dispersant additive
Copper (Cu)	anti-wear additive
Lead (Pb)	anti-wear additive
Magnesium (Mg)	detergent or dispersant additive
Molybdenum (Mo)	friction modifier
Phosphorus (P)	corrosion inhibitor, anti-wear additive
Silicon (Si)	anti-foaming additive
Sodium (Na)	detergent or dispersant additive
Zinc (Zn)	anti-wear or anti-oxidant additive

Mineral-based oils are blended with specific additives and are therefore suitable for particular applications. The different types of lubricants that are used in industry are:

- Automotive oils (gasoline, diesel, specialities)
- Gear and transmission oils
- Crankcase oils
- Turbine oils
- Heat treatment oils
- Heat transfer fluids
- Hydraulic oils
- Cutting oils
- Railroad oils
- Refrigeration oils
- Rust prevention oils
- Rubber processing oils
- Textile machinery oils
- Speciality oils.

7.3   Condition-based maintenance and oil analysis

The first instances of used oil analysis date back to the early 1940s, done by the railway companies in the western United States. Upon the purchase of new locomotives, technicians used simple spectrographic equipment and physical tests to monitor locomotive engine oils.

When diesel locomotives replaced steam locomotives, the oil analysis technique became a regular practice by railway companies. By the 1980s, oil analysis formed the basis of the condition-based maintenance programs by most railway companies in North America.

The American Navy adopted spectrometric techniques to monitor jet engines on their aircrafts in the mid-1950s. Around this time, Rolls Royce was also experimenting with oil analysis for monitoring their jet turbines. The concept of oil analysis began to spread and programs were implemented by the American Army and Air Force throughout the 1950s and early 1960s.

It is evident that the technology of oil analysis has been around for many years. Despite this, it remained secondary to the vibration approach. Vibration analysis remained the principal diagnostic technique of the condition-monitoring teams in many industries. The problem was that numerous condition-monitoring teams were not aware of the power of oil analysis. Also, in many cases when companies did have personnel assigned to perform oil analysis, these people did not interface with the vibration analysis condition-monitoring teams.

Eventually, the 1990s brought about a change in this negative trend. Many companies added oil analysis to vibration monitoring to improve their condition-monitoring programs. Likewise, several vibration condition-monitoring vendors began to offer oil analysis products, services and data management. With this integration, the condition-monitoring analyst can now obtain a more complete picture of the operating condition of the machinery and is hence in a better position to make effective decisions and recommendations.

Vibration and oil analyses complement each other. When used in conjunction, one can assist the other. For example, vibration analysis can detect a resonance but it is beyond the capability of oil analysis to detect resonance. Similarly, vibration analysis is not fully successful in detecting wear of oil-lubricated journal bearings, where oil analysis is very proficient both at detecting the wear and at assessing the severity of this particular defect. Once both analysis techniques highlight the same problem, the diagnosis and recommendations are rarely inaccurate.

An integration of the two analysis techniques in a nuclear plant was presented in a paper called 'Integration of Lubrication and Vibration Analysis Technologies' by Bryan Johnson and Howard Maxwell from the Pale Verde Nuclear Generating station. It highlights the strengths of the two techniques and the importance of them being used in conjunction. It describes the effectiveness of each technique in a tabulated form and how their combined effect is much greater. The table is shown below.

Condition	Oil Program	Vibration Program	Correlation
Oil-lubricated anti-friction bearings	Strong	Strong	Lubrication analysis will /can detect an infant failure condition. Vibration provides strong late failure stage information
Oil-lubricated journal/thrust bearings	Strong	Mixed	Wear debris will generate in the oil prior to a rub or looseness condition
Machine unbalance	Not applicable	Strong	Vibration program can detect an unbalance condition. Lube analysis will eventually see the effect of increased bearing load

Condition	Oil Program	Vibration Program	Correlation
Water in oil	Strong	Not applicable	Water can lead to a rapid failure. It is unlikely that a random monthly vibration scan would detect the abnormality
Greased bearings	Mixed	Strong	It makes economic sense to rely on vibration monitoring for routine greased bearing analysis. Many lube labs do not have enough experience with greased bearings to provide reliable information
Greased motor-operated valves	Mixed	Weak	Actuators are important machinery in the nuclear industry. Grease samples can be readily tested, but it can be difficult to obtain a representative sample. It can be hard to find these valves operating, making it difficult to monitor with vibration techniques
Shaft cracks	Not applicable	Strong	Vibration analysis can be very effective to monitor a cracked shaft
Gear wear	Strong	Strong	Vibration techniques can link a defect to a particular gear. Lube analysis can predict the type of failure mode
Alignment	Not applicable	strong	Vibration program can detect a misalignment condition. Lube analysis will eventually see the effect of increased/improper bearing load
Lubricant condition monitoring	Strong	Not applicable	The lubricant can be a significant cause of failure
Resonance	Not applicable	Strong	Vibration program can detect a resonance condition. Lube analysis will eventually see the effect
Root cause analysis	Strong	Strong	Best when both programs work together

Thus, depending on the machine and the type of defective condition that is expected, oil analysis could be implemented as the only condition-monitoring tool, or used in conjunction with the vibration analysis program, or may not be considered at all for some applications.

7.4 Setting up an oil analysis program

A systematic approach is required to set up an oil analysis program in a production plant. It consists of basically four steps:

1. Equipment audit
2. Lubricant audit
3. Monitoring
4. Program evaluation.

7.4.1 Equipment audit

The first step is to conduct an *equipment audit*. An equipment audit is a study undertaken to identify the equipment that should be monitored with oil analysis. The study investigates the mechanical and operational aspects of the equipment in detail and then determines the appropriate oil analysis approach. It sets the limit and target levels, which form a basis for maintenance scheduling.

The equipment audit includes the following.

Equipment criticality

Safety, environmental considerations, downtime cost, maintenance repair costs and machine history determine the inclusion of machines in the list of critical equipment selected for this program.

Equipment component and system identification

This involves gathering all the information regarding the machinery and understanding its complexity.

Operating parameters

This step defines the operating window of the machine. It includes the flow, pressure and temperature limits.

Operating equipment evaluation

This is a visual inspection of the machine and the aim is to identify components such as breathers, coolers and filters. Operating temperatures and pressures, duty cycle times, direction of rotation, speeds, filter indicators and related factors must be recorded.

Operating environment

Hostile environments or environmental contamination can influence lubricant degradation, eventually resulting in damaged equipment. Environmental conditions such as mean temperature, humidity and any possible contaminants must be recorded.

Maintenance history

Previous machine failures due to wear and lubrication problems must be known. This is required to set new targets and limits. These limits should be sensitive enough to forewarn failures.

Oil sampling location

The sampling location should be at a convenient position, which allows safe and easy collection of samples under normal operating conditions. The sample should have a representative concentration of wear particles in order to provide accurate information of the machine condition.

Oil tests

The physical oil test encompasses four categories:

1. Oil physical properties
2. Oil chemical properties
3. Oil contamination
4. Detection of wear particles.

Equipment-specific testing produces the required data to effectively monitor and trend the health of the lubricant and the equipment. Exception tests can verify the root cause of a change in the lubricant.

New oil baseline

A sample of the new lubricant is required to provide a starting point for the physical and chemical properties of the lubricant. The properties of oils change with operating conditions and time. To take these changes into account, the lubricant targets and alarms should be adjusted accordingly.

Targets and alarms

Original equipment manufacturer's (OEM) limits and standards can be used to set target and alarm levels. On many occasions, limits are determined from previous experience, and this experience is the best method to determine changes occurring in machinery health or lubricant condition.

Database development

A database should be developed to organize the equipment information and the collected oil analysis results, along with the target and alarm levels for the specific piece of machinery. This database should be user-friendly and must highlight machinery conditions with ease.

7.4.2   Lubricant audit

Most machines require a lubricant with the correct physical and chemical properties for reliable operation. Just like machines require regular health monitoring, lubricants must also be sampled and checked at regular intervals throughout their life cycle to ensure that they are able to meet the expected functional requirements.

Lubricant requirements

The equipment audit mentioned before will provide the information regarding the required type of lubricant specified by the equipment manufacturer. This step involves a test to check if the lubricant is within specification, such as the relationship of viscosity to operating temperature.

Lubricant supplier

Lubricants should be obtained from reputed suppliers in order to ensure consistency in supplied quality. Sampling and testing of lubricants should be done to check the quality of the supply.

Oil storage

Organized storage methods, such as labeling, are very important. A first-in first-out policy for new supplies should be realized with lubricant usage. This reduces the possibility of degradation of the lubricant properties over a period of time.

Handling and dispensing

Handling and dispensing methods have to be designed in such a way that there is no contamination, mixing or wastage of lubricants. Drain-outs, top-ups and replacements of lubricants should be recorded.

Waste oil

When a lubricant has reached the end of its life cycle, it must be disposed of in a safe way. The type of lubricant must be identified, properly labeled and the label must indicate the method of disposal. Prolonged storage of old oils must be avoided.

Baseline readings

Baseline readings are parameters of healthy equipment and lubricants taken in a short span of time. Normally three reading taken in consecutive months are sufficient to form a base for comparison with future trends.

Equipment evaluation

An analyzed sample of a lubricant can provide only partial information, unless it is accompanied by the associated machine data. This can indicate the reasons for deviations and possible root causes for any changes.

Sampling

A sampling procedure identifies the method and apparatus for collecting samples. The intention is to obtain a sensitive sample that is in accordance with changes occurring in the machine and its lubricant.

Testing

This defines the type of tests that must be conducted to provide results connected to machinery health.

Exception testing

Whenever an oil analysis detects an anomaly in any of the measured parameters, it should be verified with an *exception test*. This is an additional test to ensure repeatability and analyze from a diagnostic point of view.

Data entry, review and reporting

The results of all the samples must be logged into a data management system. These should be periodically reviewed for emerging trends and also compared with the baseline data collected earlier.

Regular reports should be generated that includes a list of recommendations. The report should mention the frequency of testing and any improvements necessary to bring the present condition of the lubricant and/or the operating conditions within acceptable targets.

7.4.3   Monitoring

Monitoring is the process of collecting data and trending the machine and lubricant conditions. This information acts as a guide for maintenance activities that ultimately lead to a safe, reliable and cost-effective operation of the plant equipment.

Routine monitoring

Routine monitoring is a fixed schedule that defines the intervals for data collection to determine machinery health.

Routes

A route is a sequence of machines aligned for sample collection. The route sequence must enable safe and non-obtrusive data collection in the shortest possible time.

Frequency of monitoring

The frequency of monitoring is dependent on the kind of equipment and lubricant. However, it can be changed as the program matures or a degrading condition is observed.

Tests

Tests form the very basis of any oil analysis program, which is to determine the present condition of the equipment. When tests indicate anomalies, immediate action must be taken in the form of exception tests, which are diagnostic in nature. It is also recommended to follow up with another diagnostic technique to confirm abnormal trends. Results of routine tests are normally available after 48 h; however in urgent cases it is possible to obtain results within 24 h.

Post overhaul testing

After the overhaul of equipment or replacement of key components, it is essential to conduct certain oil tests to ensure that the problem was rectified. Furthermore, it is important to obtain new baseline values and detect possible infant mortality conditions.

Data analysis

The data collected over a period of time from oil analysis and other allied techniques help to generate a wealth of knowledge about specific machinery. The machine operators and shop floor personnel also have valuable experience with the equipment. Their knowledge, along with the known operating conditions of the machine, forms the basis to resolve any problems that may arise. At times, repetitive problems with a seemingly unknown cause can also be resolved with the collected information and thus arrive at the root cause of failures.

Reports

All completed routes, exception testing and root cause analysis should be reported and recorded by the condition-monitoring specialist. The specialist should outline the anomaly identified and the corrective actions required. These reports should be filed in an organized way for future reference. The reports should include:

- Specific equipment identification
- Date of sample
- Date of report
- Analyst's name
- Present condition of equipment and the lubricant
- Recommendations
- Sample test result data
- Special remarks.

Use of a computerized system will allow the reports to be designed in any way required and, in many cases, will provide a machine condition overview report.

Program evaluation

Condition monitoring or predictive maintenance programs' prime function is to detect inception of failures that can influence safety, production targets and maintenance budgets. Well-managed programs are adept to perform this function. The program's success rate can be measured in terms of the number of machines monitored and the number of breakdowns prevented. However, there is usually little effort required to translate the gain from these programs into financial terms.

It is only when financial gains are demonstrated that such programs get continued support and are not curtailed or terminated as a cost-cutting measure. Dedication to an oil analysis program requires documenting all the cost benefits associated with a properly implemented program.

7.5   Oil analysis – sampling methods

One of the key elements of any oil analysis program is to obtain a sample that is not contaminated by external contaminants, 'garbage in is garbage out'. It must be appreciated that an oil analysis is searching for microscopic wear particles, and external contaminants can influence the results when being observed under a microscope.

Thus, the method of sampling, apparatus, accessories, procedure and sampling frequency all determine the informative content of the sample and this subsequently determines how beneficial results will be. The key factors to be addressed while designing a sampling program are:

- Best location – in many cases it is machine specific
- Best tools – vacuum pumps, sampling bottles, valves, traps, etc.
- Minimising contamination.

7.5.1   Sampling port location

Diagnosis of component health using oil analysis is greatly enhanced by providing sampling ports at proper locations. Ports should be in locations so that oil from individual components can be collected. This provides an analytical edge for detecting incipient failures and arriving at the root cause of failures. Sample ports are generally divided into two categories, primary and secondary (Figure 7.1).

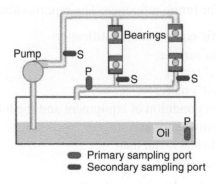

Figure 7.1
Location of sampling ports

Primary sampling ports

Routine oil samples are collected from this port. The aim is detection of wear particles, oil contamination and to test the physical and chemical properties of the oil. As mentioned before, these often depend on the type of machine and can also vary from system to system. However, the primary ports are typically located on a single return line upstream of the sump or reservoir.

Secondary sampling ports

These are sampling ports located in order to collect samples downstream of individual components. They are placed in the lubrication system to isolate upstream components. Consider the following example. An oil pump supplies oil to two bearings. The oil drained from the bearings is collected to a single return line just before it enters the sump. The primary sampling port would be located on this drain line. However, the secondary sampling ports would be located downstream of the oil pump and on the individual oil drain lines of the bearings.

As long as the sample from the primary port does not show a deteriorating trend, the secondary port samples are not required. When an anomaly is detected, samples from the secondary ports can be used to determine the source of the problem. The frequency of sampling from the problematic port alone can be increased to keep a watch on the deteriorating condition of the component.

The primary port indicates what will be going into the filters. Secondary sampling ports show what is coming out of the filters. Therefore, secondary sample ports can also be used to monitor the performance of the filters. This procedure enables a filter change based on its condition, which can be long before the differential pressure indicator shows the filter is in bypass!

7.5.2   Sampling – best tools

Drop-tube sampling

Drop-tube vacuum sampling is one of the simplest and most inexpensive methods to collect samples for oil analysis. However, one has to be careful when using this method of sampling (Figures 7.2 and 7.3).

Figure 7.2
Sampling bottle with vacuum pump

Figure 7.3
Drop-tube sampling method

When collecting the sample, the machine must be opened up, and therefore the oil is exposed to the environment. Opening a machine can potentially cause significant amounts of air-borne contamination to enter the oil and cause damage.

The key to an effective oil analysis program is the ability to draw an oil sample from a specific location while the machine is in operation and under normal load. The drop tube method, when used on equipment such as gearboxes, presents several concerns:

- The plastic tubing may be pulled into the gearbox. This is a safety hazard for the person taking the sample
- Requires a large flushing volume
- Difficult to obtain a consistent sample from the same location
- Problematic with sampling high-viscosity fluids.

Drain-port sampling

As mentioned, the best location to obtain the oil sample from a sump or a reservoir is to draw it from a location closest to the return line, gears or bearings. For circulating systems, the return line is the most ideal location. When collecting from the sump, the oil should be collected at half the depth in the sump. Generally, sumps or reservoirs are large and provide a good settling time. This allows the contaminants to settle at the bottom, and thus the bottom drain plug is not a good sampling position. Though it may be quite convenient it is still not reliable even if large volumes have been flushed.

If the drain port is the only way to obtain a sample from the gearbox, there are commercially available sample tubes that can be installed on the bottom or on the side of the sump as shown in Figure 7.4.

Figure 7.4
Commercial sample tube attached to a gearbox

These inward pilot tubes can be placed in a position so the sample is drawn from the most suitable location inside the sump or reservoir. It is important that the sample is collected from the same location inside the system at all times. This method is a more consistent and representative way of sampling oil than drop-tube sampling.

Sampling valves

There are several commercially available sample valves, which offer many different features (Figure 7.5).

Figure 7.5
Commercially available sampling valves (Minimess Valves – Hydrotechnik)

These special sample valves are similar to a check valve. The valve is normally closed until the sample port adaptor is screwed or pushed on. High-quality sample ports have a dust cap with an o-ring for second-stage leak protection. The adaptor has a hose barb on one side that accepts standard plastic tubing. Because the adaptor is screwed into the sample port it unseats the check ball in the valve and allows fluid to flow through. These valves can be used on unpressurized flooded lines and pressurized lines up to 35 MPa (5000 psi).

On pressurized systems ranging over 13 MPa (2000 psi), sampling should be conducted with a measure of safety. It is advised to use handheld pressure-reducing valves, which can be used with sample ports and adaptors to reduce pressures of 35 MPa to less than 0.35 MPa (50 psi). Another advantage of this type of sampling method is that a very small volume of static oil is retained. As a result, less oil is flushed before a sample is taken.

Figure 7.6
Procedure to collect oil sample from sampling valves (Hydra-Check Valves)

Another type of an in-line sampling valve is shown in Figure 7.6. Figures 7.6 and 7.7 demonstrate oil sampling using special sampling valves. Notice that the sampling point is at or near the inlet side of the return line filter. The four-step procedure is described below:

1. Remove the metal dust-cap.
2. Turn the spring-loaded knob 90° and flush the valve. When you release the knob it will close automatically.
3. Once again, turn the knob and take the sample.
4. Release the knob and replace the metal dust-cap.

Figure 7.7
Hydra-Check® in-line oil sampling valve

Trap pipe adaptors

These are used when oil must be sampled from vertical pipes and are basically for non-flooded type of applications. In such pipes, the oil usually spirals along the wall of the pipes. The trap pipe temporarily holds a small volume of oil in the form of a dam. From this dam, a data-rich sample is collected from the pipe (Figure 7.8).

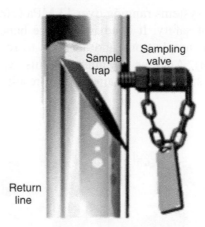

Figure 7.8
Trap pipe adaptors

Oil sample bottles

Oil analysis labs provide oil sample bottles. The repeated issue of cleanliness applies to the sample bottle as well. The ISO 3722 standard provides guidelines for cleanliness of sampling methods. This standard is used for the cleanliness levels of the sample bottle as well. Cleanliness levels are categorized as follows:

- Clean – less than 100 particles greater than 10 microns/ml
- Super clean – less than 10 particles greater than 10 microns/ml
- Ultra clean – glass bottles washed and dried in 'clean' environment.

Generally, bottles that fall under the 'clean' category are chosen. There is no point in choosing 'ultra-clean' because as soon as the bottle is opened outside the lab, the 'ultra-clean' aspect is lost. Sterilized or sanitized bottles are of no significance to oil analysis. These only ensure the absence of bacteria and are not associated with oil analysis.

A 'clean' bottle would have 100 particles >10 microns/ml. For example, if the required cleanliness code is ISO 19/16 or ISO 12/9, the ISO code specifies (for the 10-micron size range):

- ISO 19/16–1200 particles > 10 microns/ml
- ISO 12/9–9 particles > 10 microns/ml.

Apparently, the 'clean' bottle seems inadequate for the ISO 12/9 specifications. There also exists a doubt whether the *clean* bottle is adequate for the ISO 19/16 cleanliness code. To address this problem, a parameter called the signal to noise ratio (SNR) is defined.

$$SNR = \frac{\text{target oil cleanliness level}}{\text{maximum allowable bottle contamination}}$$

For the ISO 19/16 specification, the target oil cleanliness level is 1200 particles > 10 microns/ml. For a clean bottle, the maximum allowable bottle contamination is 100 particles > 10 microns/ml. Thus, the SNR in this case would be 12:1.

For the ISO 12/9 specification, the target oil cleanliness is 9 particles > 10 microns/ml. For the clean bottle, the maximum allowable bottle contamination is 100 particles > 10 microns/ml. The SNR in this case would be approximately 1:10.

Good practices indicate that an SNR of at least 5:1 should be achieved. Lower than this can cause a lot of contamination in the sample leading to erroneous readings. An SNR of 5:1 means that the noise (original particles in the bottle) can influence the particle count precision by roughly 20%. Thus, the clean bottle conforms to ISO 19/16 but is unacceptable for ISO 12/9.

As the SNR becomes higher, the bottle contamination has an insignificant influence on the particle count trends. However, if a cleaner target is required, it becomes more difficult to maintain high SNRs. Sample bottles are generally made of polyethylene terpthalate (PET), high-density polyethylene (HDPE) and glass. Each of these provides a large fluid compatibility. The preferred volumetric size is 100 ml (or 4 oz). The PET and glass bottles are clear, but HDPE are opaque and do allow an instant visual examination of the sample. Glass bottles can provide extreme cleanliness but require very careful handling. Depending on the requirement, an appropriate sample bottle must be chosen.

Sample port identification

An oil analysis program should incorporate the practice of labeling sampling ports with corrosion-resistant tags. These tags should display the information required by the technician to obtain a proper sample. The items that must be included are:

- Sample port ID (identification)
- Machine ID

- Lubricant ID
- Target cleanliness level.

Bar-coded identification tags are also a fine way to label ports.
ISO 4406 Standard for Particle Counting

Number of Particles per ml		Scale Number
More Than	**Up to and Including**	
80 000	160 000	24
40 000	80 000	23
20 000	40 000	22
10 000	20 000	21
5 000	10 000	20
2 500	5 000	19
1 300	2 500	18
640	1 300	17
320	640	16
160	320	15
80	160	14
40	80	13
20	40	12
10	20	11
5	10	10
2.5	5	9
1.3	2.5	8
0.64	1.3	7
0.32	0.64	6
0.16	0.32	5
0.08	0.16	4
0.04	0.08	3
0.02	0.04	2
0.01	0.02	1
0.005	0.01	0
0.0025	0.005	0

The ISO 4406 code specifies either two or three size ranges. The scale number refers to the number of particles per 1 ml in each size range. The scale numbers are reported with a backslash between them, with the >5 (or >2) micron scale number always first. Refer to the examples below:

2-code particle count

Size	Count	Scale
>5	3600	19
>15	28	12

ISO code = 19/12

3-code particle count

Size	Count	Scale
2	11893	21
>5	3620	19
>15	28	12

ISO code = 21/19/12

The ISO particle contamination code suggests reporting of particle count data by converting the particle count results into classes or codes. An increase from one code number to the next indicates a doubling of the particle contamination level. The ISO code scale numbers are based on total particles equal to or greater than a given size range.

Size ranges were recently updated to 4, 6 and 14 microns. Previous ranges still in use are 2 (unofficial), 5 and 15 microns. Laboratories may report counts per milliliter of sample or counts per 100 milliliters. Also, counts in each range may be reported as 'cumulative' (total particles equal to or larger than the range) or 'differential' (only within one range and the next).

For example, with the 5/15 code system, a 2-range, 100 milliliter cumulative count with an ISO code of 14/12 would represent a sample having greater than 8000 up to 16 000 particles that are 5 microns or larger in size. In this sample, there would be greater than 2000 up to 4000 particles that are 15 microns or larger in size.

7.5.3   Minimising sample contamination

The oil sample must be protected from contamination by particles in the atmosphere during and after collection. Considerable care is required to prevent 'contaminating the contaminant' in the oil sample. If atmospheric elements contaminate the oil sample, it becomes extremely difficult to distinguish the contaminants. The oil analysis program can get hampered to a great extent due to this problem.

There are many techniques to minimize contamination. These include:

- Certified bottle cleanliness
- Probe-tube bottle attachments (bottle caps are never removed)
- Ample sampling valve flushing
- Frequent cleaning/flushing of portable sampling devices (e.g., drop-tube vacuum samplers).

A slight deviation in any of the above-mentioned procedures could compromise the integrity of an oil analysis program. There are many inventive ways of collecting oil samples and prevent external contamination. One method is depicted in Figures 7.9a–c. These pictures display how the external environment is not allowed to come into contact with the bottle contents.

7.5.4   Summarising oil sampling

The sampling process is a key part of the oil analysis program. It plays a very important role in the effectiveness of the program. It demands extreme care and a systematic approach. It should include procedures and guidelines, such as:

- Frequency of sampling
- Amount of oil to be flushed before drawing the sample

- Method of obtaining the oil sample
- Tools to be used for sampling
- How to label the sample
- Routine and requirement of exception tests
- Specific remarks that could enhance improvement in reliability of equipment and integrity of the sample.

Experts believe that the time and money spent on ensuring the quality of an oil sample in the bottle pay back many times over when the benefits of oil analysis are reviewed at the end of the day. Ultimately, it is best accomplished through properly training all the people involved in oil-sampling activities. All the benefits could be lost despite all the care if an unclean pair of hands collects the sample!

(a)

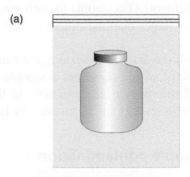

1. Place the clean bottle in a zip lock sandwich plastic bag.
2. This is done in indoors in clean room.
3. Place all these in a bigger bag along with plastic-wrapped vaccum pumps and other equipment.

Figure 7.9(a)

(b)

1. Flush the sample port/valve.
2. Without unzipping, the bottle cap is unscrewed and made to fall off.
3. Within the bag, upturn the bottle (base of bottle is at the zip end).

Figure 7.9(b)

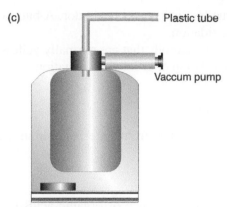

(c)   Plastic tube

Vaccum pump

1. Thread the bottle to the vaccum pump.
2. Plastic tube punctures the bag
   (turn the bottle not the pump).
3. If required make hole for
   air escape.
4. Collect right quantity of sample.
5. Unscrew pump and manipulate
   to screw back the bottle cap with
   the bag still unzipped.
6. Once sealed remove bottle from
   plastic bag.
7. Dispose plastic bag.

Figure 7.9
Collection of oil samples w/o external contamination

7.6   Oil analysis – lubricant properties

Oil analysis is primarily a combination of two types of analyses. One is of the lubricant itself and the other is analysing the contaminants in it. The tests of the lubricant are primarily to detect any degradation of the oil. Some contaminants are internally generated and some contaminants are from the environment, and this is what the second set of tests focuses on. In this section, we concentrate on the analysis of the lubricant itself. Oil is essentially a chemical fluid and is defined through a number of physical and chemical properties. Not all these properties are of interest if we are only concerned with the lubricating aspect of oils.

   The following section contains an exhaustive list and descriptions of tests that are conducted on oils and the possible information that can be derived from them. The results from these tests provide valuable information with respect to how the condition of the oil is likely to affect the function of the equipment.

7.6.1   Appearance, color, odor

This is the most elementary and basic test and is accomplished through physical impressions:

   - The oil could be clear or hazy. Haziness or cloudiness indicates the presence of water.
   - Suspended impurities could indicate wear.
   - Foaming could indicate churning or loss of anti-foaming agents.

- Usually oils have a bland odor. A burnt or pungent odor could be an indication of oxidation.
- Bright clear oils that are normally yellowish in color could appear dark red due to oxidation or gross contamination.

7.6.2 Acidity inorganic

This is to test for freedom from water-soluble strong mineral acids.

7.6.3 Acidity organic

This test detects the presence of organic acids (water insoluble but soluble in alcohol). Acids may be present as added components or may have formed during oxidation.

7.6.4 Acidity total

This is the sum of the inorganic and organic acids. The result of this test is referred to as *neutralisation value* or *total acid number* (TAN).

The TAN indicates the ability of the oil to react with the basic reagent. The test measures the acidity level of a system. The acidity level of systems increases with time, and this increase can be detected with this test. In addition, some forms of contamination can also increase the acidity levels.

To determine the TAN, a certain quantity of oil is diluted with a neutralized mixture of benzene and an alcohol-containing indicator like phenolphthalein. This mixture is then titrated against an aqueous deci-normal solution of potassium hydroxide. The result is given as the number of milligrams of KOH required to neutralize 1 g of oil. The TAN is thus a number expressed as mg KOH/gm-oil. The increase in the TAN is usually slow initially but the subsequent rise can be quite rapid.

7.6.5 Specific gravity

Specific gravity of oil is the ratio of its density with respect to water. It is used only as a means to identify the oil when its origin is in doubt. When it is suspected that the oil has been grossly adulterated with lighter or heavier products, this test is done. Otherwise, this property has little relevance in relation to the performance of the lubricant.

7.6.6 Kinematic viscosity

Kinematic viscosity is probably the most significant property of any mineral oil used as a lubricant. Viscosity is the measure of the oil's resistance to flow. Normally, any oil used for a specific application will have a viscosity range governed by the specifications of the application.

Viscosity can either increase or decrease. An increase in viscosity is due to oxidation of the oil. Oxidation of oil is an unavoidable, continuous process, but the rate of increase is determined by:

- Rate of contact with air
- Higher operating temperature
- Catalytic effect in the presence of metals like copper, iron, lead, zinc and moisture.

When an increase in viscosity occurs due to oxidation, it is accompanied by an increase in acidity. This is because oxidation of oil results in the formation of acids. When the increase in viscosity is higher than 5% of the normal value, oxidation is the probable cause. However, it may be still usable with almost a 10% increase from its rated value. A decrease in viscosity can be due to:

- Gross contamination with oil of lower viscosity
- Dilution with a light hydrocarbon like fuel
- Addition of viscosity index improvers.

When a decrease in viscosity of oil occurs due to dilution with a hydrocarbon, a drop in its flash point accompanies it. Kinematic viscosity for oil is typically specified at two temperatures, 40 and 100°C. The unit for kinematic viscosity is cSt (centistokes).

7.6.7   Viscosity index

Viscosity index (VI) is the relationship between viscosity and temperature for a fluid. It is defined as the ratio of change in viscosity due to a change in temperature in the oil. The viscosity of lubricating oil will change with change in temperature. The rate of change depends on the composition of the oil. Naphthenic-based oils change more than paraffinic-based oils. Certain synthetic lubricants change less than paraffinic oils.

A low VI means a relatively large viscosity change with temperature, and a high VI denotes a smaller change of viscosity with temperature. Hence, the VI of oil is of importance in applications where an appreciable change in temperature of the lubricating oil could influence the start-up or operating characteristics of the equipment. Machinery on a ship's deck and certain emergency equipment are examples of such cases.

7.6.8   Flash point

Flash point is the minimum temperature at which a fluid will maintain instantaneous combustion (a flash) but it is prior to continuous burning (which is the fire point). Flash point is an important indicator of fire and explosion hazards associated with a petroleum product. In used oils, it is associated with dilution, in which case the flash point is lower. At high temperatures, cracking of oil sometimes occurs, and this may also lower the flash point.

7.6.9   Pour point

Pour point is an indicator of the ability of an oil or distillate fuel to flow freely at low temperatures. It is the lowest temperature at which the fluid will flow when cooled under prescribed conditions. In many oils, it is possible to lower the pour point by using pour point depressant additives.

7.6.10   Copper strip corrosion test

A qualitative measure of the tendency of a petroleum product to corrode pure copper is determined through this test. Copper is very sensitive against sulphur-based corrosive compounds.

7.6.11   Foaming

When air is blown through enriched, contaminated or oxidized oils, foaming is observed. This test is conducted to check if the foaming is within limits. Foamy oils malfunction as lubricants. Combination of air and oil can be compared to a sponge. It is not capable of taking up the load of journals in bearings. To suppress foam, anti-foaming agents like silicones are added.

7.6.12   Saponification number

The saponification number is the measure of KOH required to completely saponify oil. Saponification number is expressed as mg KOH/gm of oil. Presence of acids, fats and oxidized products is indicated by this test. The saponification number is therefore an indication of the amount of fatty saponifiable material in compounded oil. Caution must be taken when interpreting test results if certain substances, such as sulphur compound or halogens, are present in the oil. These also react with KOH, which tends to increase the saponification number.

7.6.13   Rust-preventing characteristics

Many oils like turbine and hydraulic oils are topped with additives to prevent rusting of steel in the presence of water. This test is conducted to determine if rust prevention is normal or impaired.

7.6.14   Demulsification number

This is a measure of how quick water will separate from the oil. The demulsification number is the time it will take (in seconds) for oil to separate again after it was emulsified under specified conditions. It is a common test for turbine oils, but is sensitive to aging.

7.6.15   Oxidation tests

A number of oxidation tests are conducted to determine and judge the stability characteristics of in-service oils. These also help in determining the efficiency of anti-oxidants used in the oils. The indicators are measured in terms of acidity, viscosity, carbon residue increase or an increase in the sludge value. It also measures corrosion products and total oxidation products.

Some common oxidation tests are given below.

Oxidation stability by rotating bomb (D-2272)

It is used to assess oxidation test life in turbine or other oils. The remaining oxidation test life in used oils can be determined by measuring oxygen uptake by the pressure drop during the bomb test.

Oxidation test of inhibit turbine oils (D-943)

It is used to assess oxidation test life in turbine or other oils. It is also used to check the remaining service life in used oils.

7.6.16   Pump wear test (D-2282)

This test determines anti-wear properties of hydraulic fluids by means of a Vickers-Detroit (or similar) vane pump under standardized pressure conditions for a fixed duration. Weight loss of working parts of the pump and reduction in fluid flow are determined.

7.6.17   Emulsion and demulsibility characteristics

This test measures how quick water will separate from oil under standard conditions. Oil is emulsified with water under standardized mechanical agitation conditions, and the time to separate the oil and water is recorded. This test is conducted predominantly on turbine and hydraulic oils.

7.6.18   Air release value

This method gives a measure of the ability of the oil to separate entrained air. It is commonly applied to hydraulic and turbine oils. The air release value is defined as the number of minutes for air in oil to reduce in volume to 0.2% under test conditions.

7.6.19   Seal compatibility

This method determines the compatibility of petroleum oils with nitrile rubber seal materials in terms of change in volume and shore hardness. During this test, the initial volume and hardness of the nitrile rubber specimen is determined. The specimens are then immersed in oil at 80 °C for 100 h and subsequently cooled to room temperature. The volume and shore hardness are measured again and the difference is recorded.

7.6.20   FZG test (FZG-Niemon EP test)

This method utilizes an EP gear test rig. The purpose of this test is to determine load limits, scuffing and scoring limits on tooth flanks and weight changes in gear wheels with stepwise increased loads under standard conditions.

7.6.21   pH value

The pH value is a measure of the hydrogen ion concentration and indicates whether the fluid is acidic, neutral or basic in nature. The pH measurements are sometimes used for quality control but are not significant from a condition monitoring point of view.

7.6.22   Water content

New lubricants are devoid of water. During service, they can come in contact with water from coolers or labyrinths in steam turbines. Water makes an oil appear hazy. The crackle test (sprinkling a few drops of oil on a hotplate at a temperature of 120 °C) can also detect the presence of water. This test is sensitive when the water content is higher than 100 ppm. If the water content is high enough the oil will froth and spatter. Very small traces of water can be detected using the Karl Fischer method or the Dean and Stark method.

Karl Fischer – ASTM D-1744

This test measures water content as low as 50 parts per million, which is 0.005%. It is used in turbine system analysis, servo systems and any other system that has a low water tolerance.

Dean and Stark

Oil is heated under reflux with a water immiscible solvent like xylene or toluene between 100 and 120 °C. The solvent co-distills with the water. This vapor is then condensed in a trap. The immiscibility of the two helps in determining the quantity of water extracted from the oil.

7.6.23   Fire-resistant hydraulic characteristics

Fire-resistant hydraulic oils are oil-in-water or water-in-oil emulsions. Water glycols or non-aqueous synthetic fluids are used whenever there is a fire risk around hydraulic systems. Fire-resistant characteristics are an important property in these cases. These properties are determined by three tests:

1. Autogenous ignition temperature
2. Spray ignition temperature
3. Flame propagation test.

This is done in a mixture of fluid and coal dust.

Autogenous ignition temperature

The objective of this test is to determine the lowest autogenous ignition temperature of a hydraulic fluid at atmospheric pressure.

Temperature pressure spray ignition test

The object of this test is to determine the flammability of a hydraulic fluid, when the fluid is sprayed over three different sources of ignition.

Flame propagation test in a mixture of fluid and coal dust

The propagation of a flame is measured in a mixture of 75 g of coal and 37.5 ml of fluid. The test is conducted in an enclosure at ambient temperature without artificial ventilation.

7.6.24   Ash sulphated

This test determines the ash content of oil. The oil is first charred after which the residue is treated with sulphuric acid and evaporated to dryness. The ash content is expressed as a percentage by mass. This value determines the nature and quantity of organo-metallic additives. It is used for quality control of lubricants containing bearing ash or as a test for contamination.

7.6.25   Electrical strength

The electrical strength test is a measure of the electrical insulating properties of oils. It is a mean breakdown voltage under specified conditions. This property is easily affected by contamination and traces of water, moisture, oxidized material and fibrous material. In many cases, it can be upgraded by filtering the oil under hot vacuum conditions.

7.6.26   Specific resistance

The specific resistance is the ratio of DC potential gradient in volts/cm paralleling the current flow within the specimen to the current density in Ampere/cm^2 at a given instant of time under prescribed conditions. This is numerically equal to the resistance between opposite faces of a cubic centimeter of liquid. It is expressed in ohms centimeter.

7.6.27 Dielectric dissipation factor

The dielectric dissipation factor is the tangent of the angle by which the phase difference between the applied voltage and resulting current deviates from $\pi/2$ radians, when the dielectric of the capacitor consists exclusively of the insulating oil. These properties determine the cleanliness of the insulating oil. These properties are also related to the aging characteristics of the insulating oil.

7.6.28 Interfacial tension

It is the force necessary to detach a planar ring of platinum wire from the surface of a liquid that has a higher surface tension than the water–oil surface. It is expressed in newton/meter (N/m). To determine the extent of oxidation of used oil, for example used transformer oil, this test is very useful. This value decreases sufficiently with time during usage.

7.6.29 Extreme pressure properties

Extreme pressure (EP) additives form a protective layer on metal parts by decomposition and absorption. The anti-wear additives function in moderate temperature and pressure environments while EP additives are effective in the more extreme environments.

Molybdenum disulphide and graphite additives are a special form of anti-wear additives known as anti-seize agents. They form a protective layer on the metal parts by depositing the graphite or molybdenum disulphide. Anti-seize agents operate independent of temperature and pressure. Typical applications include engine oils, transmission fluids power-steering fluids and tractor hydraulic fluids. EP additives are common in gear oils, metal-working fluids and some hydraulic fluids.

Timken OK value

This test determines EP characteristics of lubricants and the safe load up to which no scoring on a metal test block will occur.

Four ball method (formerly called as the mean hertz load)

This test rig determines the seizure and/or scarring of steel ball surfaces under a heavy load.

7.6.30 Insolubles (pentane and hexane)

This is a test for contaminants in used lubricating oils. The oil is first diluted with pentane, causing the oil to lose its solvency for certain oxidation resins and also causing the precipitation of extraneous materials like dirt, soot and wear metals. These contaminants are called pentane insolubles. The pentane insolubles are then treated with toluene, which dissolves the oxidation resins (benzene was formerly used). The remaining solids are called toluene insolubles. The difference in weight between the pentane insolubles and the toluene insolubles is called insoluble resins.

7.6.31   Total base number (TBN)

The TBN is a property usually associated with engine oils. It is defined as the oil's ability to neutralize acid. The higher the TBN, the more acid can be neutralized by it. This quality is also referred to as the alkaline reserve and is directly proportional to the amount of active detergent contained in the oil.

New engine oils typically possess TBNs from 5.0 to 15.0, depending on manufacturer and the intended service. When the oil is used, it becomes contaminated with acids and the TBN drops. Generally, TBN levels below 3.0 are considered too low and indicate that the oil should be changed. One cause of TBN depletion is related to the use of low-quality, high sulphur fuel. During the combustion process, this sulphur turns to sulphuric acid and in turn accelerates TBN depletion. Overheating and over-extended drain intervals can cause oil oxidation. The products of oxidation are acidic and will cause the TBN to drop.

7.6.32   TAN–TBN ratio

The TAN by itself is of limited value in determining the condition of engine oil due to the fact that it represents a combination of different chemical characteristics. The acidic nature of anti-wear additives found in most modern engine oils causes a high initial TAN. The maximum benefit is acquired from the TAN by comparing it to the TBN. The TAN increases during oil service as TBN decreases. The point where these two numbers meet has been shown to be the optimal oil change interval for a certain engine under a specific load. Studies have shown that when the TAN exceeds the TBN, engine wear accelerates to abnormally high rates.

7.7   Oil analysis – contaminants in lubricants

Some contaminants are internally generated and some contaminants are from the environment. Those from the environment or outside sources include fuel, dirt, water, fuel soot and products of oxidation, nitration, sulphation or some other source. Integrally generated contaminants are particles generated by the wear of components. The analysis of such contaminants is an indication of machinery health and can form a basis of scheduling maintenance.

7.7.1   Contamination from outside sources

Contamination of lubrication oil from outside sources includes sources such as fuel, dirt, water, fuel soot, etc. This is one of the prime reasons for the oil's failure to perform its task correctly. Contamination can be detected by laboratory analysis and must be controlled by maintenance procedures. Contamination of oil from external sources can occur at any time during its life cycle. New oil is equally susceptible to this problem. Fresh oil from drums is often considered to be fresh and clean. However, the procedure of transferring new oil to the reservoir should be audited for possible sources of contamination. The likely areas to look for contamination in this sense are described below.

The oil received from the supplier is dirty. When the oil leaves the refinery, it can already be contaminated. Contamination can also occur during transportation. There are hardly any oil manufacturers that provide high cleanliness levels for oils.

On-site storage and transportation is another major source of oil contamination. Oil drums are often stored in a manner that results in contamination. Sometimes only a partial volume of oil is pumped from the drum. The pumps are then removed from this drum to another and then the drums are not capped properly. Both activities result in contamination. Not only does one type of oil gets mixed with another (even if the mixed percentage is low), but oil is also exposed to the dirty shop floor environment, water, air-borne dust and debris.

The equipment itself is another possible source of contamination. A vent, breathers, filters and seals are all possible sources of contamination. Open vents provide a passage for air-borne particles or water. Breathers and filters must be designed to keep the external contaminants away from the system. Similarly, seals that are damaged or not working properly also allow dust or water particles to enter the system. During rain or cleaning procedures, water easily makes its way into the oil system.

Oil can come in contact with fuel, glycol or coolant, especially in engines. Improper or defective equipment can cause these products to enter the system and can cause contamination, leading to adverse operating conditions. Furthermore, poor combustion may result in the oil having excessive soot levels. To prevent this kind of contamination, the following steps could be followed.

Specification for oil cleanliness may be provided to the oil supplier. The supplier may then filter the oil or may have to conduct any pre-treatment to meet the specifications. Thus, a clean supply of new oil is assured. All new oil should be stored in a clean and controlled environment. Additionally, all storage containers should be clearly marked. Containers used to transport the oil to the equipment should be clean and used for only one lubricant. The same applies for oil cans, bottles and pumps used for topping up lubricants. Vents, breathers, filters and seals should all be checked for their effectiveness in keeping contaminants out of the equipment. Machines that do not have breathers and filters should be evaluated. Breathers can be installed instead of vents, and installation of filters can provide additional protection.

7.7.2   Different external contaminants

There are many varieties of external contaminants that can adversely affect the lubrication process once it enters a system. In addition, contamination can also cause damage to the equipment.

Particles

The main external contaminant in oils is particles. Particles increase the rate of wear and oxidation. This can reduce the effectiveness of the additives in the oil. Particles traced in the lubricant are unique to the type of machine or hydraulic system. For example, in engines particles can be in the form of:

Fuel soot (diesel engines)

Soot increases viscosity, reduces the additive's ability to perform and aggravates the formation of varnish and sludge. It can be abrasive in nature causing scoring and wear. Combustion efficiency determines the extent of soot generation in engines. A poor efficiency leads to more soot formation. Soot is usually the main constituent of the total solids in diesel engines. Higher levels of soot are also attributed to improper air-to-fuel ratios, worn or stuck rings, high temperature operation, lugging and over-extended drain intervals. Infrared analysis is a method that is used to determine the amount of fuel soot present in total solid content.

Products of oxidation and nitration

The composition of oil can change due to chemical reactions like oxidation and nitration. Oil is armed with additives to inhibit such reactions. However, overheating extended drains and excessive blow by use of high sulphur fuels can deplete these additives and allow these reactions to occur. This results in excessive oil thickening, metal corrosion and lacquer formation. Whenever a severe condition is detected by testing normal solids, an infrared analysis is performed to measure the degree of oxidation and nitration.

Dirt and other environmental debris

- Dirt content is usually a small percentage of the total solids in the oil and can be quantified with a spectro-chemical analysis for silicon.
- Dirt is by far the most common cause of worn metal particles in engine oil. There is a direct relationship between the amount of dirt detected and the amount of worn metal observed. However, dirt (or any other particle) does not cause wear unless the particle is larger than the oil film separating the parts.
- It is important that the particle size and the amount of dirt should be kept as low as possible. A higher concentration of small-sized dirt particles can agglomerate and cause abrasive wear.
- The silica (also known as quartz) component of dirt causes abrasive wear. Silica, combined with metallic oxides known as silicates, is the basis of sand and nearly all rocks.
- When silica combines with carbon, it forms an abrasive called 'carborundum' which has hardness comparable to that of diamond. Silica is an oxide of silicon called silicon dioxide (SiO_2).
- Oil analysis labs do not directly examine for dirt. The silicon detected by the laboratories is from silicon dioxide. The lab readings therefore indicate the amount of silica in the oil. For simplification, silicon is another term used for dirt.
- Silicone additives used in oil for their anti-foaming properties and sand from new casting processes are other possible sources of silica.

Moisture

Water causes oxidation and rust formation, which in its turn accelerates wear. Moisture also influences the lubricating properties of the oil. Moisture enters lubricated bearing systems through different paths and becomes dissolved, suspended or free water. Dissolved and suspended water lead to rapid oxidation of the lubricant's additives and base stock. These degrade the performance of the lubricant. When water penetrates the surfaces of rolling element bearings, it can result in hydrogen embrittlement which can drastically reduce the fatigue life of the bearing. Many other moisture-induced wear and corrosion processes are common in both rolling element and journal bearings.

The best armor against moisture contamination is a three-step, proactive maintenance strategy, which includes target, exclusion and detection (TED).

Target   As a general rule, 100 ppm is a conservative contamination limit for many applications in terms of lubricant performance and bearing life. However, in certain applications, water ingress is unavoidable and in these cases a higher limit may be more practical.

Exclusion This can be achieved by more efficient use of seals and breathers in bearing housings. Flapper-valve style desiccant breathers are effective in vented systems where humid air intake and condensation/absorption are a possibility. If moisture suspension is allowed in the lubricant, water-removing filters and/or separators like centrifuges should be used.

Detection Detection is possible through the tests mentioned above or commercially available portable equipment. The lives of lubricants and machine components like bearings are largely dependent on reduced moisture levels in the oil.

Fuel contamination

Fuel contamination causes the viscosity of the oil to drastically drop (thinning of the lubricant). It dilutes the additives, increases wear and is also a fire hazard. It is clear that heavy dilution of oil is unfavorable for engines, since it reduces viscosity and also the resistance of the oil film.

The principal causes of dilution are a defective fuel injection system, a defective or choked air inlet, incomplete combustion due to a too low working temperature, badly regulated valves or insufficient compression. Dilution of used engine oil can be measured accurately by gas chromatography (GC) or by Fourier transform infrared spectroscopy (FTIR). The most common technique is to measure the flash point of oil at a certain temperature. When a flash point is detected, the dilution is heavy (more than 4%), if not, the dilution is acceptable (less than 4%).

Glycol contamination

Glycol contamination increases wear, corrosion and oxidation. Glycol is an important constituent of any coolant or anti-freeze fluid. Coolant contamination poses a serious threat to any lubrication system. Small amounts of anti-freeze can cause severe corrosion. It is especially harmful to the engine main and rod bearings where contamination can lead to engine seizure.

Glycol can enter a system from various sources. Virtually all other problems can be corrected with minimal maintenance action compared to a complete overhaul. When a glycol contamination is detected, the following should be checked:

- Leaky oil coolers
- Defective seals
- Blown head gaskets or cracked heads (engine)
- Leaking seals or gaskets on wet side liners (engine)
- Cracked block (engine)
- Contamination of new oil by dirty containers.

Detection of water/glycol leaks requires a combination of tests. The first is the water crackle test. It is a very rough test and is done by placing a drop of oil on a hot surface. The presence of water is indicated if the oil splatters or bubbles. Crackle can occur when the water percentage in the oil is more than 0.1%.

The drawback is that the detected water may be from condensation as well as coolant contamination. There is also a possibility that, even with coolant in the oil, there may be no detectable water due to evaporation at high engine temperatures. Therefore, this test should be used with caution. Water is accurately measured through the use of a Karl Fisher water tester.

The ultimate test for identifying glycol is a chemical test. The test involves shaking the oil sample with a solution of chemicals and observing the color. A positive reading in this test will be noted if there is an excess of 300 ppm (parts per million) glycol present. This level of glycol demands urgent action.

Spectro-chemical analysis has merit in detecting very small amounts of glycol. Potassium, sodium and boron are trace metals found in most coolant formulations. Their presence in engine oil can precede detection of the glycol itself. This data usually gives maintenance personnel sufficient warning that there is a pending problem.

Some caution is necessary when using this method for monitoring glycol contamination. Some oils use sodium and/or boron as part of their additive package. If these additives are used then the original oil sodium and boron levels must be considered as a baseline and amounts above these levels as contributions from glycol contamination.

Internally generated contaminants

The previous section discussed external contaminants, but there are also wear particles that are generated, washed and carried away by the lubricant as it circulates within the system. Often, the particle contamination is in itself a cause for different types of wear.

The primary types of wear are mentioned below, along with the most common cause of each. Each of these wear mechanisms results in the generation of particulate contamination capable of causing further component damage.

- Abrasive wear   particles between adjacent moving surfaces
- Adhesive wear   surface to surface contact (loss of oil film)
- Erosive wear   particles and high fluid velocity
- Fatigue wear   particle damaged surfaces subjected to repeated stress
- Corrosive wear   water or chemicals.

Abrasive wear

This form of wear results from grooving of a component surface by hard asperities or through different types of particles (rust, dust or metals) that have entered the oil. When these particles are very small, the phenomenon is known as 'abrasive erosion' (which is especially common in hydraulic systems). Introducing a filtration process can reduce abrasive wear. It is also important to ensure that vents, breathers and seals are working properly.

Adhesive wear

This occurs as a result of metal-to-metal contact, due to overheating or insufficient lubrication. This in turn causes the formation of micro-welds, with often a subsequent deposition of soft metal onto heavy metal (e.g. aluminum onto iron, lead onto steel). Consequently, there is a shearing of the junctions and a transfer of metal particles.

Cavitation

When the fluid enters a low-pressure area, bubbles are formed. However, when the fluid re-enters the high-pressure area the trapped air or gas bubbles collapse. When the implosion occurs against the surface of internal components, cracks and pits are formed. Controlling foaming characteristics of oil with an anti-foam additive can help reduce cavitation.

Corrosive wear

This is caused by a chemical or galvanic reaction. This leads to the removal of material from a component surface. Corrosion can be a direct result of acidic oxidation. A random electrical current can also cause corrosion. The presence of water or combustion products can promote corrosive wear. It can be avoided by the use of more corrosive-resistant materials and also through the use of neutralising additives in the oil and replacing the oil at proper intervals.

Another form of corrosion is due to surface contact. Removal of material occurs between two surfaces that are in almost static contact but are subjected to mechanical vibration and oscillation. This is called *contact corrosion* or *fretting corrosion*. Consequently, there is oxidation of certain particles. Thus, for iron materials, there is an accumulation of 'red powder'. In mechanical seals, wear between a secondary seal like a Teflon wedge and the shaft sleeve is an example of fretting corrosion.

Cutting wear

Cutting wear is caused when an abrasive particle has imbedded itself in a soft surface. Equipment imbalance or misalignment can contribute to cutting wear. Proper filtration and maintenance is imperative to reduce cutting wear.

Fatigue wear

This occurs when cracks develop in the component surface that leads to the generation of particles. Insufficient lubrication, lubricant contamination, vibration, high pressure, high temperature and other aggressive conditions provoke fatigue wear.

Sliding wear

This type of wear evolves during equipment stress. When machines operate at high speeds or loads, sliding wear occurs. The excess heat in an overload situation breaks the lubricant film and results in metal-to-metal contact. Sliding wear can also occur when a stationary part comes into contact with a moving part.

Electrical wear

The erosion due to electrical sparks causes this wear type. Sparks are produced by inadequate electrical insulation in motors and alternators.

7.8   Particle analysis techniques

Once contaminants enter the circulation system they tend to cause wear and failure of components. However, if these contaminants are trapped and subsequently analyzed it is possible to trace their origins and the extents of damage that can be expected. To analyze the contaminants, many methods like spectrometric analysis, infrared analysis, particle counting, wear particle analysis and others are used.

Each technique has niche applications and its own limitations. These limitations can be compensated for through the use of other techniques. This makes oil analysis a powerful predictive maintenance tool.

7.8.1   Spectrometric analysis

Spectrometric analysis is one of the main techniques used in particle analysis. It is primarily used for trending the accumulation of small wear metals and elemental constituents of additives and for identifying the possible introduction of contaminants. The results of a spectrometric analysis are typically reported in ppm (parts per million). Spectrometric analysis monitors the smaller particles present in the oil. Large wear metal particles will not be detected.

In the spectrometer, oil is excited electrically to the point where it will emit light. Each element present in the burning oil emits a light of characteristic color and frequency. Spectrometers translate the intensity of the colors into a computerized readout. A typical report from this test would list nine major wear metals for industrial gear oil and hydraulic oils. A test for twelve elements is conducted for automotive oils. The computer compares the present amount of wear metals with a fresh oil sample and also with samples from similar machines. The computer also compares the results of previous samples taken from the same equipment to establish wear trends. Below is a list of elements that can be detected using spectrometric analysis and the possible components from where they could originate.

Wear Elements	Probable Origin
Aluminum	Pistons, bearings, blower/turbo chargers, pump vanes, thrust washers and bearings, blocks, oil pump bushings, housing clutches, impellers, rotors
Chromium	Rings, roller-taper bearings, liners, exhaust valves, coolant, rods, spools, gears, shafts, anti-friction bearings
Copper	Bearings, thrust washers, bushings, oil coolers, oil additives, wrist-pin bushings, cam bushings, valve-train bushings, governor, oil pump, steering discs, pump thrust plates and pistons, injector shields, wet clutches
Iron	Cylinders, crankshafts, valve train, piston rings, clutch, pistons, rings, gears, bearings, liners, shafts, plates, blocks, camshafts, pumps, shift spools, cylinder bores and rods, piping and components of circulating oil systems
Lead	Bearings, gasoline additives, oil additives
Magnesium	Oil detergent, oil alkaline reserve
Molybdenum	Oil additive, friction modifier
Nickel	Alloy, gear plating, valve guides and ring bands, shafts, anti-friction bearings
Silver	Wrist-pin bushings, anti-friction bearings, silver solder
Tin	Bearings, piston plating, alloy of bronze (copper/tin), bushings
Silica (silicon – silicone)	Ingested dirt and sand, gasket sealant, oil anti-foam additive, anti-freeze additive
Zinc	Anti-wear oil additive (zinc dialkyl-dithio-phosphate), galvanized parts in circulating oil systems

However, every lubricant contains additives. Additives contribute to the display of elements detected with the spectrometric method. The following are some of the elements that are contained within additives. The detected additive elements can be used to monitor the consistency of the lubricant, as opposed to the additive's effectiveness.

Barium (Ba)	detergent or dispersant additive
Boron (B)	extreme-pressure additive
Calcium (Ca)	detergent or dispersant additive
Copper (Cu)	anti-wear additive
Lead (Pb)	anti-wear additive
Magnesium (Mg)	detergent or dispersant additive
Molybdenum (Mo)	friction modifier
Phosphorus (P)	corrosion inhibitor, anti-wear additive
Silicon (Si)	anti-foaming additive
Sodium (Na)	detergent or dispersant additive
Zinc (Zn)	anti-wear or anti-oxidant additive

The most common elements tracked through spectrometric analysis are iron, lead, tin and copper. Coolant contamination can be detected by monitoring the concentrations of sodium and boron. Dirt is detected by tracking silicon. The main focus of this analysis is to trend the accumulation of small particles of wear metals and elemental constituents of additives, as well as to identify the possible introduction of contaminants. Spectrometric analysis is efficient for particles less than 5 microns in size. It should be kept in mind that spectroscopy is not sensitive to detect large particles in an oil sample. Unfortunately, the large particles are more indicative of an abnormal condition or wear mode.

The most severe wear modes such as spalling, severe sliding wear and cutting wear generate large particles, which will go unnoticed through spectroscopy tests. Large contaminant particles are also overlooked by spectroscopy. The particle size at which spectrometers lose their detection capability depends on a number of factors, such as the spectrometer type and model. Nevertheless, it is generally agreed that spectrometers lose their ability to detect particles in the 5–10 micron range. To evaluate particulates larger than 5–10 microns, other methods must be considered.

For larger wear particles there are other techniques available, such as:

- Acid digestion method
- Microwave digestion method
- Direct read (DR) ferrography
- Rotrode filter spectroscopy (RFS).

The first two methods are difficult to incorporate as a predictive analysis tool due to the time they consume and their high cost. These methods provide a list of the sample's total concentration of elements. The acid digestion method can provide the ratio of large and small particles in the original sample but it is not obtained with the microwave digestion method. The third method, called DR ferrography, is sensitive to both large and small ferrous particles in lubricating oil, and it is capable of determining a ratio whereby abnormal wear could be characterized in rotating machinery.

However, DR ferrography is still based on magnetic separation, and this is the main disadvantage when the sample contains *non-ferrous* metal particles of all sizes or includes large non-metallic inorganic particles such as sand or dirt.

Rotrode filter spectroscopy (RFS) was first introduced in 1992. This spectrometric technique detects large or coarse wear metals and contaminants in a used oil sample. All particles up to 25 microns are the first indicators of abnormal wear situations. The RFS provides a low cost and efficient ferrography method and is superior to DR ferrography because it detects ferrous, non-ferrous and contaminant elements (usually 12 elements).

Detection efficiency of large material gets poorer as particle size increase above 25 μm in diameter. Its accuracy range is within 15%.

As mentioned, the atomic absorption spectroscopy (AAS) and inductively coupled plasma (ICP) techniques of spectrometric analysis are not sensitive to particles above 5 microns. Rotating disk electrode (RDE) spectrometers have a somewhat higher range but even they are limited to 10 microns.

In the RDE spectrometer, a carbon disk is pressed onto the end of a rotating shaft. A certain volume of oil is poured from the sample bottle into the sample bottle cap and positioned so that the bottom of the rotating disk passes through the oil. A spark gap is formed between the top of the rotating carbon disk and the tip of the carbon rod electrode. An electric discharge across the gap vaporizes the oil that remains stuck onto the rotating disk. The emitted light contains the wavelength characteristics of the elements in the oil sample. The spectrometer optics and electronics quantify these wavelengths and can report up to 20 elements in 30 or 40 s in ppm. The carbon discs are known as rotrodes.

Subsequent innovations paved the way for a new technique called rotrode filter spectroscopy (RFS) that can analyze larger wear particles. The RFS makes use of the fact that the carbon disk electrodes used in RDE are porous in nature. A fixture is used to clamp the discs (Figure 7.10). The oil samples are drawn from the outer circumference of the disk electrodes when a vacuum is applied inside the discs.

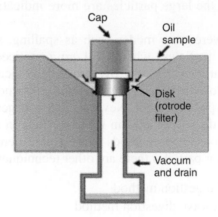

Figure 7.10
RFS sample preparation fixture

The oil is then washed away with a solvent, the disk is allowed to dry and the particles are left on the outer periphery so they can be vaporized and detected when sparked on the RDE spectroscope. In practice, an oil analysis should be conducted by first using the RDE technique, which provides an analysis of dissolved and fine wear particles. A second analysis using RFS should then be performed to quantify the larger particles. The two tests provide an indication of the wear particle size distribution in the sample. Thus, a sudden burst in larger particles is quickly detected using the RFS method.

By combining large particle results with the conventional RDE analysis of the dissolved and fine particles in the oil sample, a complete wear analysis picture can be obtained for the machine under investigation. The technique has several advantages which make it a powerful tool to monitor machine failures. Most abnormal wear modes cause a significant increase in the concentration and size of wear particles.

Rotrode filter spectroscopy is fast and efficient and is used as a standard screening test for every oil sample that enters the lab. The data are also excellent for contamination analysis because it provides the elemental composition of large contaminant particles (such as silicon). The beneficial feature of this fact is that it is possible to determine whether the element is from an additive package (such as a silicone polymer for defoaming) or a contaminant (sand/dirt particles).

7.8.2 Infrared analysis

Infrared (IR) analysis is an absorption form of spectrometric analysis. This technique is not an atomic absorption technique used for detecting specific elements, but rather detects specific combinations or groups of atoms called *functional groups*. The different functional groups assist in determining the material properties and their expected behavior.

Certain infrared wavelengths are absorbed by each functional group. Thus, an appropriate wavelength is directed at the sample being analyzed, and the amount of energy absorbed by the sample is measured. The amount of absorbed energy is an indication of the extent of presence for that particular functional group in the sample. It is hence possible to quantify the results. The unit of measurement is usually expressed in absorbency units (AU).

The presence (or absence) of functional groups and their quantities in the oil sample provide useful information about the new or used oil. It is possible to identify the physical composition of the oil, the additives in the lubricant and also possible contamination and degradation of oil.

Physical composition   Most of lubricants consist of the base stock and specific additives. Different base stocks can be distinguished using IR analysis. This might be required during quality assurance. It is also useful in detecting contamination of synthetic oils with mineral oils.

Additive chemistry   The presence and quantity of any particular additive can be determined with ease using this technique. However, an original sample of the oil is required for comparison purposes. Depletion of specific additives can be verified through the use of this technique.

Note: The amount of additive or additive depletion is measured by the amount of additive metals present in the oil. It is possible that the tests used to determine these metals could be misleading. Additive metals can easily fluctuate between +20% and −20% under normal circumstances. Many elements like copper, which is present in the additives, are also present in machine components. As a result, it is possible to have a loss of an additive metal and a secondary gain from an internal component. Consequently, it may not be possible to detect actual additive metal loss. It is also possible that a wide variety and quantity of additives were used to meet the specification for the oil. Therefore, the level of a metal detected to determine an additive does not necessarily reflect the usefulness of that additive in the oil.

Contamination   Oxidized oils can be detected using IR analysis. This technique can also detect contamination of oil by fuel, glycol or water. Sometimes, specific chemicals being manufactured by a factory find their way into lubricants. This type of contamination can also be exposed.

Lube degradation   *Oxidation* and its resulting products cause the degradation of oil. Products of oxidation are organic acids. When oxidation becomes excessive, it tends to corrode the surfaces of equipment components. *Nitration* of oil is also possible and is common in natural gas-fueled engines where fixation of nitrogen can occur. Defects such as varnishing, sludge deposits, sticky rings, lacquering and filter plugging occur in systems with oxidation and/or nitration problems. *Sulphation* occurs as a result of a 'blow-by' of combustion products into the oil. These combustion products (and the fuel itself) contain sulphur. The sulphur reacts to form acidic products, which can build up and cause severe corrosion. The formation of these products is called sulphation. This is also a cause for the degradation of oil.

Fourier transform-infrared analysis

The IR analysis was first introduced in 1979. This initial method was replaced several years later by an instrument that is much more efficient, accurate and easy to use called the Fourier transform-infrared analysis (FT-IR). With this technique, a beam of light is focused through a film of used oil and the wavelengths are then compared to light transmitted through new oil of the same type.

It is good practice to collect samples from new oil deliveries to establish baseline information. The used oils can consequently always be referenced with baseline readings. The differences in readings provide information with respect to the degradation of the used oil. The differences or changes of interest in FT-IR are soot, oxidation, sulphur products and nitration. The FT-IR instrument is also used to detect the presence of water, glycol (anti-freeze) and fuel. If the likelihood of water, glycol or fuel is indicated, separate physical confirmation tests for that contaminant may be required.

This technique is also occasionally used to determine the optimum life of the lubricant. Initially, the oil is frequently tested at short intervals, like 50 h. Once the probable life of the lubricant is established, the new frequency for testing or optimal replacement time of the oil can be fixed. The IR analysis is the most accepted method used in industry today to determine the condition of lubricants. The method investigates the additives in the oil and the actual condition of the oil itself. This technique offers a true and complete analysis of the lubricant in a way that no other test can duplicate.

Particle counting

Particle counting is a technique that is used in conjunction with wear metal analysis. Wear metal analysis cannot measure all particles. Hence, excessive particulate contamination can be overlooked if particle counting is not a routine part of the oil analysis program. The information obtained is critical to determine if excessive wear or dirt entry is occurring. There are many methods for analysing the distribution of particle sizes. These are based on different principles and are listed in the table below.

Particle counting tracks all particles within a sample that typically ranges in size from 5 to 100 microns (at times up to 200 microns). However, particle counting does not differentiate the composition of the materials present. All particles are counted and reported over a particle size range.

Type of Measurement	Characteristic Feature	Measurement Technique
Optical	Imaging light extinction light scattering diffraction	Microscopes HIAC and ROYCO ROYCO (gas) Laser
Electrical	Resistance	Coulter
Geometric	Screen	Sieving
Gravitational	Sedimentation	Andreason
Cyclonic	Inertia	Donaldson Alpine
Magnetic	Wear debris	Ferrography

In one of the particle-counting techniques, light from a solid-state laser is used to illuminate the sensing area on the fluid. The light scattered by particulates passing through this intensive beam is collected at right angles with respect to the beam and focused onto a photo detector. All pulses over a pre-selected discrimination level are converted into a digital format. Once in digital format, the pulses can be counted to yield an indication of the size of the particle(s). In another technique, the particle counter operates by shining a laser beam through the oil. Particles block the laser and prevent them from being detected by special detectors.

This device has several limitations. It cannot measure dark oils (e.g. soot-containing engine oil), severely contaminated oils, or samples that contain water. This test is sometimes utilized in oil cleanliness testing and is performed on all non-engine samples.

7.8.3 Wear particle analysis/ferrography

Wear particle analysis (WPA) or ferrography is a technology that utilizes microscopic analysis to identify the composition of materials. This test method evaluates the particle type, shape, size and quantity of the particles. Ferrography is a non-invasive examination of the oil-wetted components of machinery. The particle's size, shape and composition allow a process of elimination in which the abnormal wear of specific components can be identified. Using ferrographic techniques, it is possible to identify wear-related failures at an early stage, before collateral damage or failure occurs. Wear particle analysis is used in two ways.

In the first method, a routine monitoring and trending of the solid content in the lubricant sample is prepared. The particle size, shape and quantity are indicative of the machine condition. A healthy machine rarely ejects particles greater than 10 microns in size. However, as deterioration sets in, the size and number of the particles also increase.

The second method involves observing and analysing the type of wear (as described in Section 7.6.3) from the shape of the particle. Analysis of particles with this technique involves their separation from the lubricant in the following way. A lubricant sample is diluted with a solvent like tetrachloroethylene (TCE) and allowed to flow down a specially prepared low-gradient inclined slide while passing across a bipolar magnetic field.

The force that attracts the particles is proportional to their volume, whereas the viscous resistance of the particles to motion is proportional to their surface area. The flow rate of the process is such that non-ferrous particles and contaminants are randomly deposited due to gravity down the length of the slide substrate. After this process, a solvent is used to remove the lubricant remaining on the slide. This deposition of particles on the slide is called a *ferrogram* (Figure 7.11). When the ferrogram has dried, the wear particles and solid contaminants are stuck on the slide surface and are ready for examination under a microscope.

A 3× magnification biochromatic microscope is used to identify and examine the particles remaining on the slide. Particles caused by known wear modes have distinctive characteristics, which reveal the wear mechanism at work. Analysts can classify wear particles through size, shape, concentration and metallurgy. Various types of particulate contamination can also be identified. Classification and identification of particles help to assess the machine's wear condition and make appropriate recommendations.

Once the ferrogram is made, heat treatment and acid/base solution introduction are two common methods used in the identification of individual particles. Heat treatment is a process by which the glass slide is exposed to a controlled heat source for a specified length of time. By heating it in stages, the effects of oxidation due to the heat source are observed. A change in surface definition of the particles can be an indicator of their composition. Color is another significant aspect in the morphology of particles.

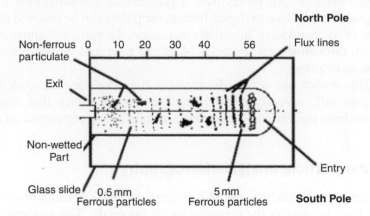

Figure 7.11
Ferrogram

Acid/base solution introduction is a particularly useful method when dealing with non-ferrous materials. These have different melting temperatures than the ferrous particles, and the addition of an acid or base solution to the slide can help identify the particles through the observed reactions. Another aspect of analytical ferrography is the identification of the wear mechanism or the type of relative motion that created the particles. Particles are not only created through component surfaces in contact. Contaminants contained in the lubricating fluids also come into contact with component surfaces, generating very specific wear patterns.

Given below are brief descriptions of some very common types of wear. These cover almost 95% of all the possible wear patterns that can be observed in practice. Figures 7.12–7.29 and their descriptions were sourced from a ferrogram analysis chart prepared by Predict DLI, Cleveland Ohio, USA.

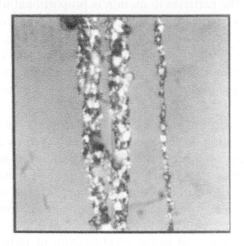

Figure 7.12
Normal rubbing wear

Normal rubbing wear

These are flat platelets less than 15 microns in major dimension. They are considered as normal machine wear (Figure 7.12).

Severe sliding wear

Flat, elongated particles greater than 20 microns with striations. They are indicative of excessive load/speed on the sliding surface (Figure 7.13).

Figure 7.13
Severe sliding wear

Cutting wear

Long, curled strips of metal. These are indicative of misalignment or abrasive contamination in the lubricant (Figure 7.14).

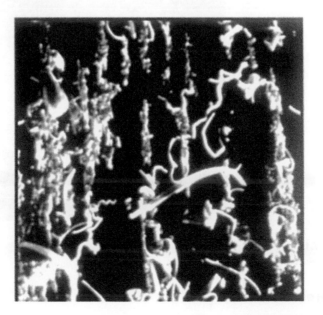

Figure 7.14
Cutting wear

Gear wear

Flat striated particles observed with 1000× magnification. They are caused by fatigue, scuffing or scoring of gear teeth (Figure 7.15).

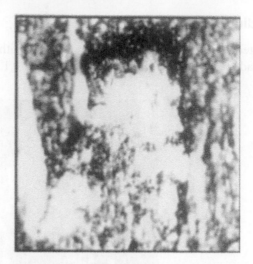

Figure 7.15
Gear wear

Bearing wear

Laminar platelets, as depicted in Figure 7.16, are visible with 1000× magnification. These are indicative of rolling contact failure.

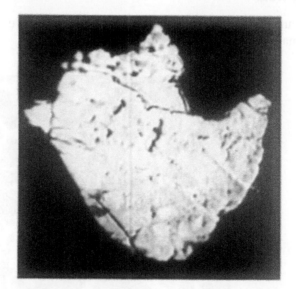

Figure 7.16
Bearing wear

Spheres

Small spheres (less than 5 microns) with a magnification factor of 1000×. This is an early warning of rolling element-bearing failure (Figure 7.17).

Figure 7.17
Spheres

Black oxides

Black particles aligned in a magnetic field with a 1000× magnification. This is an indication of insufficient lubrication (Figure 7.18).

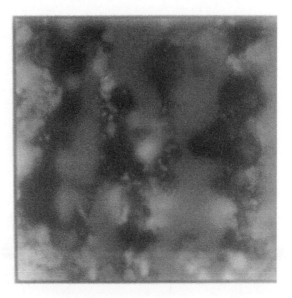

Figure 7.18
Black oxides

Red oxides

Red-orange particles aligned in the magnetic field with a 1000× magnification factor. These are caused by water in the oil or poor lubricant condition (Figure 7.19).

Figure 7.19
Red oxides

Corrosive wear

With a magnification factor of 100×, heavy concentration of fine particles is detected at the exit of the ferrogram. This indicates depletion of oil additives (Figure 7.20).

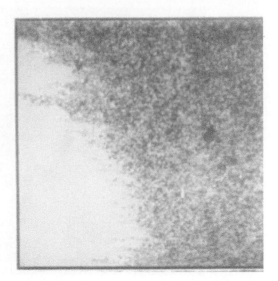

Figure 7.20
Corrosive wear

Aluminum wear

A 500× magnification shows a white metal particle misaligned with respect to the magnetic field. It is indicative of wear of an aluminum component (Figure 7.21).

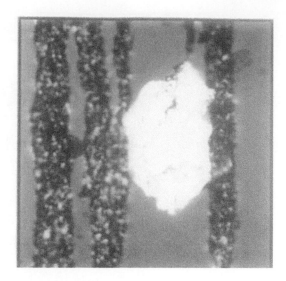

Figure 7.21
Aluminum wear

Copper alloy wear

Gold-colored particles not aligned with the magnetic field. This is due to wear of brass or bronze components. The magnification is 100× (Figure 7.22).

Figure 7.22
Copper alloy wear

Dust/dirt

Foreign particles of material, not characteristic of machine or oil. These are usually sand or dirt (Figure 7.23).

Figure 7.23
Dust/dirt

Fibers

With a 100× magnification, fibers are visible as non-aligned and passing transmitted light. It indicated failure of filter media in this case (Figure 7.24).

Figure 7.24
Fibers

Friction polymer

The picture shown in Figure 7.25 was taken through a red filter with 500× magnification. Amorphous materials are visible that pass transmitted light. This indicated excessive stress or load on the lubricant.

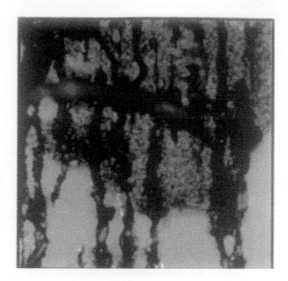

Figure 7.25
Friction polymer

Break-in wear

Under 1000× magnification, long and thin bar-shaped particles are visible. These are generally observed with new machinery and is caused by so-called break-in wear (Figure 7.26).

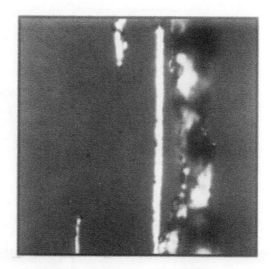

Figure 7.26
Break-in wear

Case-hardened and low-alloy steel particles

The picture shown in Figure 7.27 was taken with a magnification of 400×. It is the result from a heat-treated (at 330 °C) ferrogram. It shows purple- and blue-tempered colors, which indicate case-hardened and low-alloy steels, respectively. This is usually gathered from gears with abnormal wear conditions.

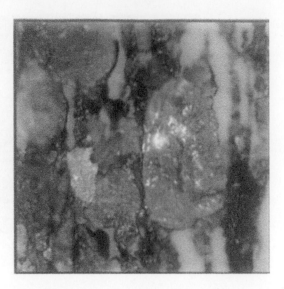

Figure 7.27
Case-hardened and low-alloy steel particles

Lead/tin babbitt

Figure 7.28 shows a non-ferrous particle before and after heat treatment, with a magnification of 500×. Babbitt is an indication of journal bearing wear.

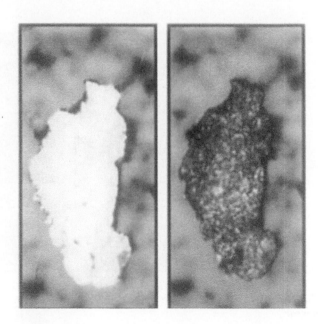

Figure 7.28
Lead/tin babbitt

Molybdenum disulphide

A non-ferrous particle, gray in color with many shear planes. It is a solid lubricant additive in the system. The magnification is 400× (Figure 7.29).

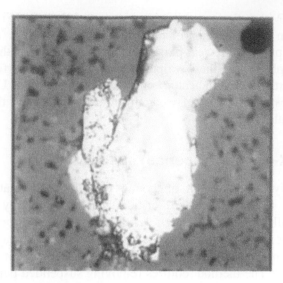

Figure 7.29
Molybdenum disulphide

DR ferrography

In the previous section, we have seen that magnetic wear particles are separated from the oil and deposited on a glass slide. The pattern of particles obtained on the slide is known as a ferrogram. Microscopic examination permits characterisation of the wear mode and probable sources of wear in the machine. This technique is the wear particle analysis and is also known as *analytical ferrography*. An automated version of this magnetic separation technique is direct read (DR) ferrography.

The DR ferrography measures the ratios of large and small particles in the sample, and the data are used to calculate the wear particle concentration and severity index .

These two parameters can be used for trending and are excellent indicators of abnormal, ferrous wear, but are unsuitable for non-ferrous wear. The test is very valuable when a wear trend has been established. In any machine, at least one of two surfaces in contact is ferrous. Brass bevel gears mesh with steel worn gears. Plain bearings are always in contact with a ferrous shaft. Although the non-ferrous surface would wear first, the corresponding ferrous wear can be detected and monitored through DR ferrography.

One method of DR ferrography is through optical emission of light sources to sort and count ferrous particles. The laboratory equipment would typically house two separate light sources. The prepared sample travels through a glass precipitator tube that rests on a large magnet. The magnet in turn traps the ferrous particles, and a numerical value representative of the percentage of light blocked is produced. This 'density' value is reported as the DL or *density of large* particles (over 5 microns in size) and DS or *density of small* particles (under 5 microns in size). With these two values, a *wear particle concentration* (WPC) is calculated as follows:

$$WPC = DL + DS$$

WPC is trended over a period of time to evaluate the rate of wear. Any changes can be observed by the increase or decrease of the WPC value. In addition to the WPC, a *percentage of large particles* (PLP) or a *severity index* can also be calculated using the density values:

$$PLP = \frac{DL - DS}{WPC} \times 100$$

The PLP is an indicator of the severity of the wear. The occurrence of abnormal wear results in the generation of particles above 5 microns. Thus, determining the percentage of large particles in wear particle analysis can identify the necessity to carry out analytical ferrography. Both these values, namely the wear particle count and the percentage of large particles, are indicators of machinery condition. Unfortunately they do not provide any clues to the cause of wear. In order to determine the root cause, one must revert to analytical ferrography.

7.8.4   XRF (X-ray fluorescence) spectroscopy

Oil analysis consists of analysing wear particles that are generated during the wear process. However, modern critical machines employ high-efficiency filters that trap almost all the large wear particles, leaving very little for the analyst. Hence, we must shift the focus to the filter where the wear data might be trapped. The debris deposited in the filters can reveal the machine's health. This information is typically lost when the filter is discarded. The method requires removing suspended particles from the circulating oil and analysing their constituent chemical elements by X-ray fluorescence (XRF) spectroscopy. The debris deposited in the filters reveals the machine's state of health, as well as the amount of wear particles generated between two filter changes.

The XRF spectroscopy, like other spectrometric techniques, entails the excitation of electrons from their orbits (Figure 7.30). This leads to emission of UV rays with characteristic frequencies, which can be analyzed. During Rotrode atomic emission spectroscopy, an electrical discharge produces plasma, causing thermal emission (similar to the extreme temperatures of an argon plasma torch). When the atoms return to the normal state, the excess energy is emitted as light. Each element emits light at different frequencies on the electromagnetic spectrum. The amount of light emitted at a given frequency corresponds to the concentration of the element present in the sample.

The XRF works in a similar way. The difference is that XRF excites the atoms with a bombardment of X-rays. In return, the atoms emit X-rays that are characteristic of the chemical element, at an amplitude that corresponds to the mass of the chemical element in the sample. The XRF spectrometers are calibrated to report the concentrations of the important elements present in parts per million (ppm).

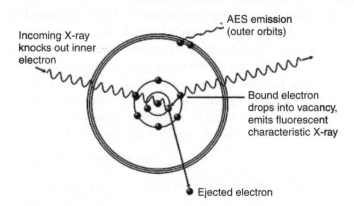

Figure 7.30
XRF spectrometers

The XRF excites electrons within the inner shells of the atom, near the nucleus. The AES excites the electrons in the outer shell, which results in the emission of visible light.

In a Rotrode AES spectrometer, the particles are excited by plasma. This plasma is generated via an electrical arc in a process that is limited to measuring particles less than or equal to 10 microns.

The inner shell excitation, which occurs in an XRF spectrometer, is brought about by high-energy X-ray emissions. The XRF is thus capable of measuring fine particles, as well as large metal samples (e.g. metal plates). The method is, however, limited by the depth from which X-rays can enter and exit the sample material. In the case of a metal plate, a thin sample will yield the same results as a thicker sample. This leads to a particle size effect dependent on the X-ray penetration depth. Despite this, XRF still produces large signals for large particles and small signals for small particles.

7.9 Alarm limits for various machines (source: National Tribology Services)

Compressors

Oil Analysis Test	Typical Allowed
Spectroscopy – iron	15 ppm
Spectroscopy – copper	500 ppm
Spectroscopy – lead	15 ppm
Spectroscopy – aluminum	15 ppm
Spectroscopy – chromium	15 ppm
Spectroscopy – tin	15 ppm
Spectroscopy – zinc	500* ppm min
Spectroscopy – nickel	15 ppm
Chlorine	20 ppm
Viscosity	+20 to −10% of nominal ISO grade
Water	500 ppm max
TAN	1.0 mg KOH/g max

* This applies to the additive package in the lubricant and is a minimum.

Turbines

Oil Analysis Test	Typical Allowed
Spectroscopy – iron	15 ppm
Spectroscopy – copper	500 ppm
Spectroscopy – lead	15 ppm
Spectroscopy – aluminum	15 ppm
Spectroscopy – chromium	15 ppm
Spectroscopy – tin	15 ppm
Spectroscopy – zinc	500 ppm min.
Spectroscopy – nickel	15 ppm
Chlorine	20 ppm
Viscosity	+20 to −10% of nominal ISO grade
Water	500 ppm max
TAN	1.0 mg KOH/g max

Hydraulic systems

Oil Analysis Test	Alarm Limits	Further Action by Lab
Spectrochemical silicon	15 ppm	
Spectrochemical copper	12 ppm	
Spectrochemical iron	26 ppm	
RFS (doublecheck)	2:1 ratio or greater coarse to fine	Ferrography
Viscosity	+20%, −10% of nominal ISO grade	
Oxidation	0.4 Abs/0.1 mm over last sample	TAN (1.5 mg KOH/g max)
Particle count*	17/14 ISO code	
Water	0.1% max	Karl Fisher

*Particle count for many hydraulic systems is much lower.

Gas-fired engines

Oil Analysis Test	Alarm Limits	Further Action by Lab
Spectrochemical analysis	10% increase over last sample	
RFS (doublecheck)	2:1 ratio or greater coarse to fine	Compare to trend ferrography
Viscosity	+20 or −10% of nominal ISO grade	
Oxidation	0.2 Abs/0.1 mm over last sample max 1.00	TAN
Nitration	0.2 Abs/0.1 mm over last sample max 1.00	TAN
TBN	Half of new oil value or equal to TAN 1.0 mg KOH/g min value	

7.10   Conclusion

In the past, companies considered, the vibration analysis program as the mainstay of the predictive maintenance philosophy. However, with technical improvements, sophisticated laboratories and portable equipment, oil and wear particle analysis is starting to yield benefits and financial returns. The method is now rapidly growing as a desirable predictive maintenance technique that should be complementary to vibration analysis.

Implementation of an oil analysis program can significantly reduce maintenance costs for a production plant and can also improve plant reliability and safety. Oil analysis used for condition monitoring should have a baseline established. When joined with significant historical data, it provides very valuable information about machinery health.

The costs of a properly implemented program should be recoverable through the extension of lubricant change intervals. Increased reliability, availability, the prevention of unanticipated failures and less downtime are added benefits.

8

Other predictive maintenance techniques

8.1   Introduction

There are dozens of predictive maintenance technologies, and some have become a standard in many industries. The 'standard' technologies include vibration analysis, ultrasound, oil analysis, wear particle analysis and thermography. In the earlier topics, we discussed in detail the predictive maintenance techniques of vibration analysis and oil and wear particle analysis. In this chapter, we will discuss the other predictive maintenance techniques like ultrasound and thermography. These techniques are quite useful for some applications and complement the major techniques very well.

Many condition-monitoring techniques can be used to monitor the same fault condition. For example, a problem identified by an oil sample can also be cross-checked with vibration analysis or thermography. An electrical problem, identified with ultrasound, can be confirmed through the use of thermography. Hence, a confirmation of the diagnosis is possible through the use of the different predictive maintenance techniques. This fact makes the predictive maintenance more convincing, especially when critical machinery is involved.

8.2   Ultrasound

Most machines emit consistent sound patterns under normal operating conditions. These sound patterns (sonic signatures) can be defined and recognized, and therefore changes in these signatures can be seen as components begin to wear or deteriorate. In a certain sense, it is a different form of mechanical vibrations. Sound is a microscopic oscillation at the molecular level of a substance. Vibration is a macroscopic oscillation of structures – in other words, physical structures that move. Ultrasound is defined as sound waves that have frequency levels above 20 kHz, higher than the range of human hearing. Air-borne ultrasounds operate in the lower ultrasonic spectrum of 20–100 kHz and have the following properties:

- Small objecots easily block air-borne ultrasound.
- It does not penetrate solid surfaces (though it can go through cracks).
- Air-borne ultrasound radiates in a straight line.
- Ultrasound does not travel a great distance.

The above properties of ultrasound are exploited to make it a useful technique in providing early information in cases of:

- Leak detection in pressure and vacuum systems (e.g. boilers, heat exchangers, condensers, chillers, distillation columns, vacuum furnaces, speciality gas systems)
- Bearing inspection
- Steam trap inspection
- Valve blow-by
- Integrity of seals and gaskets in tanks, pipe systems and large walk-in boxes
- Pump cavitations
- Detection of corona in switch gear
- Compressor valve analysis.

8.2.1 Ultrasonic translator

The ultrasonic translator (Figure 8.1) is generally a lightweight, handheld device that can be easily carried within the plant and also into confined spaces.

Figure 8.1
An ultrasonic detector

Air-borne ultrasound translators are relatively simple to use. They consist of a basic handheld unit with headphones, a meter, a sensitivity adjustment and (most often) interchangeable modules that are used in either a scanning mode or a contact mode. Some instruments have the ability to adjust the frequency response between 20 and 100 kHz. An ultrasonic transmitter called a tone generator is often also included.

The ultrasound can be 'heard' after it has been modified and processed into the audible range by a process called heterodyning. Heterodyning, as the word implies, is the mixing of two waves. The mixing of two waves produces the sum and the difference of the original waves, which also allow the shifting of a high-frequency sound to the audible range. The mathematical formula used for this process is:

$$A\cos(\omega_1 t) \cdot B\cos(\omega_1 t) = 1/2\left[AB\{\cos(\omega_1 + \omega_2)\}t\right]$$
$$+ 1/2\left[AB\{\cos(\omega_1 - \omega_2)\}t\right]$$

where A is the amplitude of the first wave, B the amplitude of the second wave, ω_1 the frequency of the first wave, ω_2 the frequency of the second wave.

As an example, let us assume that we have a bearing generating a sound signal of 31–33 kHz. Since humans can only hear up to about 16–20 kHz, this sound cannot be heard. Mixing a 30 kHz constant frequency wave with this signal, we will get a difference of 1–3 kHz, and when added a total of 61–63 kHz. The 1–3 kHz information can be heard and interpreted. The sum is much higher, out of the audible range, and is discarded.

As seen in Figure 8.2, the frequency from the oscillator must be adjusted to be close to the input signal. However, there are also some instruments that analyze a broad spectrum of ultrasound and select the proper frequency band of the signal automatically. Because their operation is like that of an AM radio, manually adjusted heterodyne ultrasonic detectors can miss key signals if they are not set properly. The automatic option in some instruments eliminates the guesswork associated with manual frequency adjusting. It translates sounds from 20 to 100 kHz to the audible range, regardless of the frequency of the original signal.

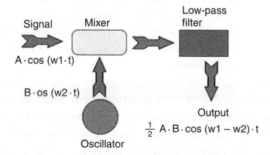

Figure 8.2
Heterodyning

The intensity of the ultrasound can be determined by observing a meter on the instrument. Some instruments include the ability to adjust the intensity of the signal. For example, if an ultrasound source is too difficult to locate due to an intense signal, the user can focus on the exact position by adjusting the sensitivity downward. On the contrary, when a low-level leak occurs in a water valve, the frequency tuning can be adjusted to help a user hear the trickle of the water leak.

The interchangeable modes allow users to adjust the instrument for different types of inspections. The scanning mode is used to detect ultrasounds that travel through the atmosphere, such as a pressure leak or a corona discharge, while the contact mode is used to detect ultrasounds generated within a casing such as in a bearing, pump, valve or steam trap housing.

8.2.2   Ultrasound detection techniques

Air-borne ultrasound detection begins with the setting up of the detector. It is initially kept on the logarithmic setting on the meter selection dial. A 'fixed band' position must be selected on the frequency selection dial. The sensitivity selection is kept to maximum. Scan is initiated by pointing the module towards the test area. The procedure is to go from the 'gross' to the 'fine' with more subtle adjustments made as the leak is approached. During this operation, high background noise as well as low-intensity signals could be encountered. There are techniques that can be used to overcome this which are described in the following section.

8.2.3   Isolating competing ultrasounds

Leak detection using the ultrasound technique is an easy task. An area is scanned to detect for a distinct rushing sound. With continued sensitivity adjustments, the area is scanned until the loudest point is heard. In some instruments, a rubber-focusing probe is provided. This attachment narrows the area of reception to pinpoint a small emission. The rubber-focusing probe is also an excellent tool for confirming the location of a leak. This is done by pressing it against the surface of the suspected area. The leak is confirmed if the sound of the leak remains consistent. If it decreases in volume, the leak is elsewhere. At times, competing ultrasounds make it difficult to isolate a leak. In such a case, the following could be tried.

If possible, switch off the equipment with the suspected leak. Otherwise, isolate the area with a physical barrier. Manipulate the instrument and use shielding techniques.

Manipulation of the instrument is achieved by getting as close as possible to the suspected leak and pointing away from the competing ultrasound. Isolate the leak area by reducing the sensitivity and by pushing the tip of the focusing probe to the suspected area, checking small sections at a time. Since ultrasound occurs at high frequencies, it has a short wavelength, and can thus usually be blocked or shielded.

In some extreme instances, when leak detection proves difficult in the fixed band mode of the frequency selection dial, the leak sound should be tracked by 'filtering' out the problematic sounds. In such a case, adjust the frequency selection dial until the background sound is minimized and then proceed to listen for the leak.

Some of the common techniques to isolate competing ultrasounds are:

- Use your body to shield the test area to act as a barrier for the competing sounds.
- Place a clipboard close to the leak area and turn it so that it acts as a barrier between the test area and the competing sounds.
- Wrap your hand (gloved) around the rubber-focusing probe tip so that the index finger and the thumb are close to the end. Place the rest of your hand on the test area so that there is a complete barrier between the test area and the background noise. Move the instrument and your hand together over the various test zones.
- Along with the gloved hand method, a rag can be wrapped around the rubber-focusing probe tip. Care must be taken that the end of the tip does not get blocked. This is usually the most effective method since it uses three barriers: the rubber-focusing probe, the gloved hand and the rag.
- When a large area must be covered, reflective material such as welder's curtain or a drop cloth can be used to act as a barrier. The barrier is placed to act as a wall between the test area and the competing sounds. Sometimes the barrier is draped from ceiling to floor, and at other times it is hung over the railings.

Detecting weak ultrasounds

The previous discussion focused on the case when there is an excess of ultrasound in the environment and it is difficult to pinpoint the source. There are also opposite situations, especially in leak detection, when the ultrasound signal is weak. In these cases, manipulations are made to generate or enhance the intensity of the ultrasonic signal. Ultrasonic leak detection is based on the turbulence generated when a fluid exits from a narrow opening. When the turbulence is high, the ultrasound signal is proportionately loud and vice versa.

When the leak rate is low, the turbulence is hardly detectable. In these cases, the pressure in the system should be increased (if possible) to obtain a better signal. If this method is not practical, a *liquid leak amplifier* should be utilized.

Certain specially formulated liquids produce a thin film through the escaping gas passes. Under low flow conditions, small bubbles are formed. These bubbles may not be visible to the naked eye. However, when these bubbles implode, ultrasound is generated and can be picked up by the detector as crackling in the headphones. Leaks as low as 1×10^{-6} std. cc/s can be detected.

Another method involves using an ultrasonic transmitter called a *tone generator*. This is often used with equipment such as heat exchangers, tanks and vessels. The tone generator is placed inside the tank. It will generate an intense uniform ultrasound signal. The detector is placed outside. If it detects the signal, it is an indication of a leak. The preferred method for leak detection is through using pressure or vacuum, but the tone generator method is a good backup method for difficult situations.

8.2.4   Ultrasonic applications

Leak detection

During a leak, a fluid (liquid or gas) moves from high pressure to a low pressure. When the gas or liquid passes through the defect in the pipe or vessel, a turbulent flow is generated. This produces a broad range of sound. The high-frequency ultrasonic components of these sounds have extremely short wavelengths, and short wavelength signals tend to be fairly directional. The strong ultrasonic components are heard through headphones and seen as intensity increments on the ultrasound meter. It can be stated that high intensities generally cause higher ultrasound levels.

Steam and air leaks

Ultrasound detection is very efficient for leak detection, particularly for detecting steam and air leaks. These leaks can be expensive but many companies let them go unnoticed. The ultrasound, when properly detected and measured, provides the user with the location and severity of the leak. Common applications for ultrasound include leak detection for pneumatic and other gas systems, vacuum systems, gaskets and seals, and steam traps. Ultrasound can also detect valve blow-through. Since many small leaks are difficult to find simply by listening, the ultrasound technique helps to discover the many small leaks that can result in significant losses over long time periods.

Heat exchangers, boilers and condenser Leaks

The leakage paths in boilers, heat exchangers and condensers are the tube, tubesheets and housings. The ultraprobe can be used to detect leaks in three ways:

1. Pressure leaks
2. Vacuum leaks
3. Utilising the ultrasonic tone generator method.

When detecting vacuum leaks, the turbulence will occur within the vacuum chamber. For this reason, the intensity of the sound is less than that of a pressurized leak. Though the method is most effective for low, medium to gross leaks, the ease of ultrasound detection makes it useful for most vacuum leak problems. While it may be necessary to take a unit off-line to inspect for leaks, it is often possible to perform an inspection while on-line or at partial load with ultrasound.

Bearing and mechanical inspection

Ultrasound is also adept in detecting early failures in bearings. Ultrasonic inspection of bearings is useful in recognising the beginning of fatigue failure, brinelling of bearing surfaces, flooding or lack of lubrication. In ball bearings, as the metal in the raceway, roller or bearing balls begins to fatigue, a subtle deformation occurs. This deformation of the metal will produce an increase in the emission of ultrasonic sound waves.

When testing, a change in amplitude of 12–50 times the original reading is an indication of incipient bearing failure. When a reading exceeds any previous reading by 12 dB, it can be assumed that the bearing has entered the beginning of a failure mode. This applies to monitoring bearings in a frequency range of 24–50 kHz.

In the case of a lack of lubrication, the sound levels increase as the lubricant film reduces. A rise of about 8 dB above the baseline accompanied by a uniform rushing sound will indicate a lack of lubrication. To avoid overlubrication, do not lubricate if the baseline reading and baseline sound quality are maintained.

The ultrasonic frequencies detected by the system are reproduced as audible sounds. This signal can assist a user in determining bearing problems. When listening, it is recommended that the user becomes familiar with the sounds of a healthy bearing, often heard as a rushing or hissing noise. Crackling or rough sounds indicate a bearing in the failure stage. In some instances, a damaged ball emits a clicking sound, whereas a high intensity, uniform and rough sound may indicate a damaged race or uniform ball damage. Loud rushing sounds similar to the normal rushing sound of a healthy bearing, but only slightly rougher can indicate a lack of lubrication.

Monitoring slow-speed bearings is also possible. Due to the sensitivity range and the frequency tuning, it is quite possible to listen to the acoustic quality of slow-speed bearings. In extremely slow bearings (less than 25 rpm), it is often necessary to disregard the output meter and only listen to the sound of the bearing (headphones). In these extreme situations, the bearings are usually large (1–2 in. and larger) and greased with a high-viscosity lubricant.

Often, no sound will be heard, as the grease will absorb most of the acoustic energy. If a sound is heard, usually a crackling sound, it is an indication of a defect. On most slow-speed bearings, it is possible to determine a baseline level and monitor at regular intervals. It is suggested that the attenuator transfer curve method be used, since the sensitivity will usually have to be higher than normal.

Valve and steam trap inspection

Valve and steam traps have a tendency to leak. When this happens, it can prove to be quite expensive in terms of product quality, safety and energy loss. Valve operation will influence the way fluids flow through a system. Valves are usually checked for leaks with the contact probe on the downstream side. This is achieved by first reading the upstream side and adjusting the sensitivity to read about half of fullscale. The downstream side can then be measured and the sound intensity between the two compared. If the signal is significantly lower than upstream, the valve is considered to be closed. If it is louder than upstream and is accompanied by a rushing sound, it is considered to be leaking.

Steam traps can also be inspected easily with ultrasonic translators (airborne ultrasound instruments). A contact probe is again used to inspect a steam trap. Listening to the trap operation and observing the meter can interpret the trap condition. The speed and simplicity of this type of test allow every trap in a plant to be routinely inspected. Valves and steam traps work in different ways. The ultrasound detectors make it easy to adjust for these differences and readily determine operating conditions while valves and traps are operational.

Electrical inspection

The ultrasound technique can be used to detect problems in electrical apparatus such as insulators, cable, switchgear, bus bars, relays, contactors, junction boxes. In substations, components such as insulators, transformers and bushing can be inspected. Failure of these components is unacceptable in industrial plants, power transmission and distribution. As a result, they need to be monitored on a regular basis. There are basically three types of electrical problems that can be detected using ultrasound:

1. *Arcing:* An arc occurs when electricity flows through space, like lightning.
2. *Corona:* When voltage on an electrical conductor, such as an antenna or high voltage transmission line, exceeds threshold value, the air around it begins to ionise to form a blue or purple glow.
3. *Tracking:* Often referred to as 'baby arcing', it follows the path of damaged insulation.

Though ultrasound can be used to measure the electrical problems in the low, medium and high voltage systems, it is generally used in systems above 2 kV. When electricity escapes in high voltage lines or when it jumps across a gap in an electrical connection, it disturbs the air molecules around it and generates ultrasound. This sound will often be perceived as crackling or frying sound; in other situations it will be heard as a buzzing sound.

The method for detecting electric arc and corona leakage is similar to the procedure outlined in leak detection. Instead of listening for a rushing sound, a user will listen for crackling or buzzing sound. This is especially useful in identifying tracking problems. In enclosed switchgear, the frequency of tracking greatly exceeds the frequency of serious faults identified by infrared. It is recommended that both tests be used with enclosed switchgear.

On lower voltage systems, a quick scan of bus bars often will pick up a loose connection. Checking junction boxes can reveal arcing. As with leak detection, the closer one gets to the leak site, the louder the signal. If power lines are to be inspected and the signal does not appear to be intense enough to be detectable from the ground, you can use an ultrasonic waveform concentrator (a parabolic reflector), which will double the detection distance of the system and provide pinpoint detection. In addition, the problems of RFI and IVI also have an impact on communication networks. These conditions produce ultrasound and can be detected using this technique.

Compressor inspection

Compressors are central and critical to any compressed gas system. Routine inspection and maintenance can prevent unexpected downtime. Although any type of compressor can be inspected ultrasonically, the most common application is on large reciprocating compressors. Specifically, valve function in these compressors is critical. Minor valve leaks can rapidly lead to large leaks, which impacts production and plant safety.

8.3   Infrared thermography

Any object with a temperature above absolute zero emits energy. As its temperature rises, the energy emission also increases. Infrared thermography is a technique that produces a visible graph or a thermographic image of thermal energy radiated by objects. Thermography utilizes a portion of the infrared band of the electromagnetic spectrum between approximately 1 and 14 microns. This bandwidth is usually associated with infrared radiated energy produced by object temperatures of −20 °C and higher.

Electronic instruments used in infrared thermography utilize a lens system to focus the invisible energy radiated from an object's surface onto the infrared sensitive detector(s). The various energy levels are measured by the detector(s) and then transformed into a visible image with each energy level represented by a different color or gray scale level. This image can be viewed by the user on a LCD or CRT display. The image can be stored digitally or on video, for review, analysis and reporting at a later stage. All infrared systems (simple to complex) are sensitive to infrared radiated energy only. They do not actually measure temperature. They are useful in applications where a variation in temperature, reflection, surface condition or material may cause a difference in the radiated energy level that can be detected by the infrared camera.

The new generation of infrared cameras is known as focal plane arrays (FPAs). High-resolution infrared images are now possible with these commercially available instruments due to the fact that most FPAs use over 75 000 detector elements in the production of each infrared image. Radiometric FPAs use an onboard computer system to perform a series of complicated functions to calculate temperatures. These temperatures are only valid when a trained and experienced operator enters a series of accurate parameters. Thermography is a useful predictive maintenance technique in the following ways:

- It does not make contact with the surface.
- The technique does not involve any hazardous actions.
- It can be used in hazardous zones.
- It is not affected by electromagnetic waves.
- Like other predictive techniques, it is used while systems are operating.
- It provides instant information.
- Data can be collected and stored in digital format.

However, thermography has some disadvantages, such as:

- The cost of the hardware can be quite high.
- Some systems have software limitations.
- The emissivity of the object must be calculated or should be known.
- The ambience should be homogenous with respect to the thermal energy – any radiating source close to the area being monitored can affect the thermal scans.
- Distance, atmospheric conditions and temperature can affect the quality of images.

8.3.1 Applications of IR thermography

Thermography is applied for fault detection in the following areas.

Electrical equipment

- Cooling systems, earth faults, circulating currents, laminations, cracking insulation.
- Auxiliary equipment such as fuses, relay contacts, switchgear, distribution boards and transformers can suffer from loose connections, imbalanced loads or corrosion, which result in resistive heating (Figure 8.3). These conditions can be identified with thermography.
- In a three-phase system, unequal temperatures can indicate imbalance in the phases.

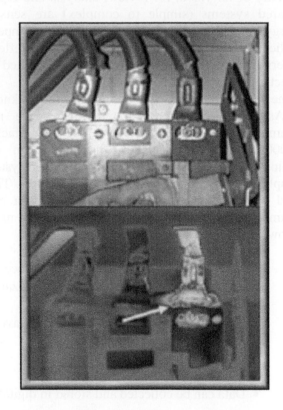

Figure 8.3
Resistive heating at an electrical connection (source: www.cinde.ca)

Thyristors are used to control the speed of large motors. When these are connected in parallel it becomes difficult to detect if one of them has a problem. It is possible that a fuse has blown, but this may not be possible to detect by the change in motor speed. Besides, it is dangerous to inspect a live electrical system. Thermography is helpful under such conditions. In high-voltage systems, thermography can be used to monitor transformers, power lines, fuses and fuse holders, overheating power factor capacitors, switchgear, control panels, isolators, circuit-breakers, relay contacts and connections.

In electrical motors (Figure 8.4), the following problems can be detected through the use of thermography:

- Problems in the cooling system of motor enclosures, poor electrical connections, overheating of the frame, overheating of rotor body and windings, lamination hot spots on stator cores, problems in stator windings, cracking of insulation at stator end windings, overheating of bearings and seals.
- In motors with commutators and brushgear, thermal images can indicate uneven wear. This can occur due to poor contact of the brushes with the commutator segments.

Figure 8.4
A hot endshield of an electric motor (source with permission: www.infraredthermography.com)

Mechanical equipment

- Misalignment or bent shafts can be detected with thermography. In the case of misalignment, frictional heat is generated by the sliding contact of gears in a gear coupling or by friction between the flexible members of other types of couplings. This results in loss of energy in the form of heat, which can be monitored with thermal imaging.
- Defective reciprocating compressor valves increase in temperature because the hot compressed gas moves back and forth across a valve. Some compressors have many valves on the cylinder. A thermal scan can very quickly detect the defective valve.
- Insufficient lubrication results in frictional heat, which can be detected with thermography.
- Damaged bearings, gears, chains, clutches, belt slippage and belt wear can be detected by using thermography.
- Linkages and actuators with excessive frictional contact can be detected.

- Leaking/passing of valves can be identified.
- Leakage detection in hydraulic systems, which could include leaks across piston and rod seals, various types of valves and valve boxes, pumps, pipes, fittings and hoses.
- Tank level measurement is possible with the use of thermography (Figure 8.5).

Figure 8.5
Notice the liquid level and the sedimentation in the tank (image with permission from: www.infraredthermography.com)

Energy systems

- Boiler and steam systems, flues, heat exchangers and regenerators. In this case, integrity of the insulation, brickwork and poor joints of furnaces and kilns can be assessed. Condition of refractory line ducts can be determined. Operation of blast furnace stoves, burners' flame patterns and water-cooled elements can be monitored with thermography (Figure 8.6).
- Regenerators, insulation, buildings, roofing can be monitored. Heating and air conditioning leakages can be traced.

Electronic systems

Discrete components, printed circuit boards (PCBs) and electronic bonding can be quality-checked with thermal images. Electronic and microelectronic components like PCBs are made of small components whose individual temperatures cannot be monitored. It is well known that higher temperatures could exponentially reduce the life of electronic components. Sophisticated thermographic systems have been designed that can automatically inspect complete PCBs by measuring their thermal content. For very small components, infrared microscopes are used for thermal monitoring (Figure 8.7).

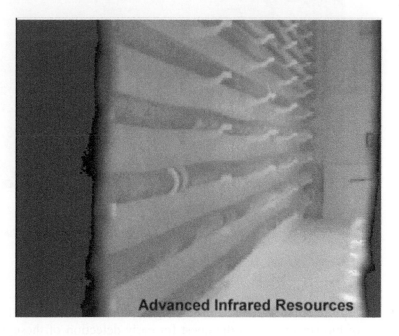

Figure 8.6
A furnace picture to inspect refractory condition (image with permission from:
www.infraredthermography.com)

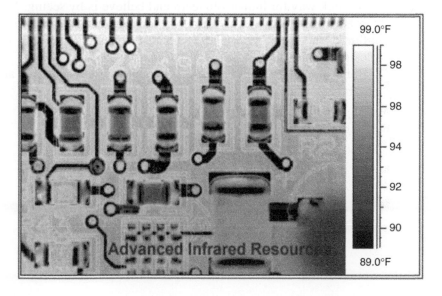

Figure 8.7 (a)

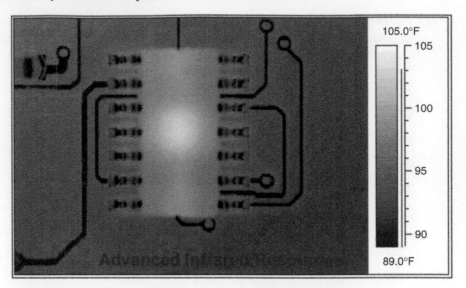

Figure 8.7 (b)
Thermographic testing of PCBs (image with permission: www.infraredthermography.com)

8.4 Conclusion

Equipment in distress sends out signals that are not within the perception range of human senses. In the quest for early detection of these signals, techniques like ultrasonics and infrared thermography evolved as additional predictive maintenance techniques.

Through the technology of ultrasonics, we can hear sounds out of the audible range. These sounds can indicate defects in machinery.

A quick way for humans to learn and believe is by seeing. More information can be transferred visually than with any other sense. This is the strength of infrared thermography. The technique generates pictures that graphically illustrate phenomena not normally visible to the human eye.

Both these techniques have their limitations, but for certain applications they are the best choice.

Appendix A
Exercises

1 Predictive maintenance basics

1.1 The maintenance philosophy of operating the machine until it fails is called:

    (a) Breakdown maintenance
    (b) Proactive maintenance
    (c) Predictive maintenance
    (d) Preventive maintenance.

1.2 The maintenance philosophy under which an operating machine is taken up for maintenance based on the condition of its health to prevent a breakdown is called:

    (a) Breakdown maintenance
    (b) Proactive maintenance
    (c) Predictive maintenance
    (d) Preventive maintenance.

1.3 Root cause failure analysis is a technique adopted under:

    (a) Breakdown maintenance
    (b) Proactive maintenance
    (c) Predictive maintenance
    (d) Preventive maintenance.

1.4 Vibration analysis is a technique adopted under:

    (a) Breakdown maintenance
    (b) Proactive maintenance
    (c) Predictive maintenance
    (d) Preventive maintenance.

1.5 Name the *three* categories under which plant equipment can be classified.

1.6 In a typical process plant, there are *five* cooling water supply pumps. Depending on the weather conditions, four of the five pumps are operational at any point in time. Electrical motors with a rating of

1.5 MW drive these pumps. Under which category would you classify these pumps?

1.7   Predictive maintenance principles are analogous to:

  (a) Crime detection
  (b) Medical science
  (c) Forensic medicine
  (d) Failure analysis of electronic components.

1.8   One of the following is *not* a predictive maintenance technique:

  (a) Root cause failure analysis
  (b) Ultrasonic thickness measurement
  (c) Performance evaluation
  (d) Vibration analysis.

1.9   Vibration analysis (in the detection mode) is done:

  (a) On a regular basis for selected machines
  (b) To detect the severity of fault in a machine
  (c) To ascertain the time after which machine or its component will fail
  (d) To detect faults in vibration measuring tools.

1.10   Vibration analysis and the trending of vibration data cannot detect one of the following:

  (a) Rotor imbalance
  (b) Fouling of internals
  (c) Bad bearings or gears
  (d) Machinery alignment.

2   Vibration basics

2.1   A spring-mass system has three characteristics that resist the vibrations caused by an external force. Tick the one that is *not* one of them.

  (a) Mass
  (b) Damping
  (c) Stiffness
  (d) Young's modulus.

2.2   System response of a body is proportional to the:

  (a) External forces like unbalance
  (b) Restraining forces
  (c) Ratio of the external force to the restraining forces
  (d) Difference between the external forces and the restraining forces.

2.3   In a spring mass system, the maximum amplitude of vibration is 5 cm. It has an angular velocity of π rad/s. If the mass is at the origin or center point at the zeroth second, what is the amplitude after 2 s?

2.4 Number of cycles per second of a wave is a measure of:

(a) Amplitude
(b) Phase
(c) Frequency
(d) Wavelength.

2.5 Amplitude of vibration is *not* a measure of:

(a) Displacement
(b) Velocity
(c) Acceleration
(d) Phase.

2.6 Phase difference is:

(a) The offset between the crests of two time waveforms
(b) The offset between the troughs of the two time waveforms
(c) The offset between similar points of the two time waveforms
(d) The angle difference of a point on waveform from the origin.

2.7 A square waveform is a resultant of:

(a) Time waveforms having odd harmonic frequencies
(b) Time waveforms having even harmonic frequencies
(c) Time waveforms having integer harmonic frequencies
(d) None of the above.

2.8 Crest factor of a waveform is a ratio of:

(a) Peak to rms
(b) rms to peak
(c) Peak to mean
(d) Peak to 1.414 times rms.

2.9 For a machine running at 9 Hz (540 cpm), which parameter is most suitable for measuring vibrations:

(a) Displacement
(b) Velocity
(c) Acceleration
(d) Phase.

2.10 In the ISO 2372 vibration standard, the acceptable vibration levels are based on:

(a) kW rating of machines
(b) Types of foundation of machines
(c) Ratio of rotor mass to casing mass
(d) Both the kW rating and type of foundation of the machine.

3   **Data acquisition**

3.1   Velocity pickups are based on the principle of:

  (a) Coil in magnet
  (b) Magnet in coil
  (c) Piezoelectric crystal
  (d) Coil in magnet or magnet in coil.

3.2   Charge mode accelerometers are used:

  (a) When very high frequency range is to be measured
  (b) When temperature of the surface where vibrations are collected is high
  (c) For low-speed applications
  (d) In hazardous areas.

3.3   Which of the following is used to measure shaft vibrations:

  (a) Photocell
  (b) Velocity transducer
  (c) Accelerometer
  (d) Eddy current probes.

3.4   In Eddy current probes, the oscillator/demodulator demodulates the signal and provides a modulated DC voltage where the DC portion is directly proportional to:

  (a) Gap between the probe and shaft
  (b) Vibration of the shaft
  (c) Mechanical and electrical runout of the shaft
  (d) Eddy currents generated by the high radio frequency.

3.5   Which of the following vibration transducers does not need an external power source:

  (a) Accelerometers
  (b) Velocity pickups
  (c) Eddy current probes
  (d) None of the above.

3.6   Measurements from handheld vibration meters can be subject to errors on account of:

  (a) Position of measurement and probe angle
  (b) Probe type and the pressure applied
  (c) All the above-mentioned reasons
  (d) None of the above.

3.7   Phase difference measurement is possible using:

  (a) Accelerometer
  (b) Data collector
  (c) Velocity pickup
  (d) Dual channel analyzer.

3.8   Data manager software collects and stores data from:

  (a) Online monitoring systems
  (b) Data collectors/analyzers
  (c) Handheld vibration meters
  (d) Torsional vibrometers.

3.9   Which of the following does not need a reference on the shaft for phase measurement:

  (a) Stroboscope
  (b) Photocell
  (c) Electromagnet/non-contact pickup
  (d) Dual channel analyzer.

3.10   Torsional vibration represents changes in:

  (a) Relative angular displacement between two points on a rotating shaft
  (b) Angular velocity of the shaft
  (c) Angular acceleration of the shaft
  (d) Mean Torque.

4   Signal processing, applications and representations

4.1   As per the Nyquist sampling theorem, the sampling rate should be at least:

  (a) Half of the highest frequency of interest
  (b) Twice of highest frequency of interest
  (c) Same as the highest frequency of interest
  (d) Three times the highest frequency of interest.

4.2   Which of the following windows provide good amplitude resolution of peaks between bins and minimal broadening of the peak:

  (a) Flat
  (b) Rectangular
  (c) Hanning
  (d) Hamming.

4.3   If the maximum frequency set on the analyzer (F-max) is 320 000 cpm and the resolution is set as 1600 lines, the bandwidth is:

  (a) 200 cpm
  (b) 0.005 cpm
  (c) 2000 cpm
  (d) 5 cpm.

4.4   Overlap helps in:

  (a) Improving frequency resolution
  (b) Improving amplitude accuracy
  (c) Enhancing processor speed
  (d) Reducing time to collect vibration data.

4.5 The *beats* frequency caused by two time waveforms with marginally different frequencies is equal to the:

(a) Average of two frequencies
(b) Summation of the two frequencies
(c) Difference of the two frequencies
(d) Ratio of the two frequencies.

4.6 When the direction of rotation is opposite to motion of shaft centerline during a revolution, the orbit thus generated would be in:

(a) Forward precession
(b) Reverse precession.

4.7 A cascade plot is a representation of:

(a) Frequency – amplitude – time
(b) Frequency – amplitude – phase
(c) Frequency – amplitude – speed
(d) Frequency – amplitude – FFT.

4.8 Technique for visualisation of vibratory movement of a machine under its operating load is called:

(a) Modal analysis
(b) Cepstrum analysis
(c) Operational deflection shape analysis
(d) Finite element method.

4.9 Coherence *cannot* provide any meaningful information in which of the following cases:

(a) Identifying source and effect of vibration
(b) Selecting the location of sensors
(c) Reducing the number of sensors
(d) Detecting a bearing defect in the high-frequency range.

4.10 In one-third octave spectral analysis the bandwidth is:

(a) Constant
(b) Proportional
(c) Exponential
(d) One-third of F-max.

5   Machinery fault diagnosis using vibration analysis

5.1 In an overhung fan rotor with an unbalance, the confirmatory test is:

(a) $1 \times$ rpm peak with similar phase readings in axial direction on both bearings
(b) $1 \times$ rpm peak with similar phase readings in radial direction on both bearings
(c) $1 \times$ rpm peak with unsteady phase in axial direction
(d) $1 \times$ rpm peak with opposite phase readings in axial direction on both bearings.

5.2 Axial phase readings with sensors on each bearing of a bent shaft will have:

(a) No phase difference
(b) Phase difference of 180°
(c) Phase difference of 90°
(d) Unsteady phase readings.

5.3 If the phase difference is 180° measured in the radial direction on bearings across the coupling of a pump and a motor, the suspected fault is:

(a) Foundation looseness
(b) Damaged coupling
(c) Angular misalignment
(d) Parallel misalignment.

5.4 When sub-harmonic multiples of $^1/_2\times$ or $^1/_3\times$ are observed in the FFT spectra, it could indicate:

(a) Oil whirl
(b) Oil whip
(c) Rotating stall
(d) Internal assembly looseness.

5.5 Resonance frequency of structural members can be detected with:

(a) Phase analysis
(b) Time waveform analysis
(c) A 'bump test'
(d) Fast Fourier transform.

5.6 When two dots and two blank spaces are observed in an orbit plot, it indicates that the vibration frequency is:

(a) $^1/_2\times$ rpm
(b) $1\times$ rpm
(c) $2\times$ rpm
(d) There is no correlation.

5.7 If the assembly phase factor $N = 1$, the number of gear teeth is 98 running at 5528 rpm and the pinion has 65 teeth, what is the hunting tooth frequency?

(a) 85 cpm
(b) 255 cpm
(c) 170 cpm
(d) 541 744 cpm.

5.8 A motor running at 1450 rpm is installed with a pulley with 120 mm diameter. The belt length is 1295 mm. The pump pulley diameter is 180 mm. The belt defect frequency will be:

(a) 422 cpm
(b) 633 cpm

    (c) 107 cpm

    (d) 161 cpm.

5.9   If a rotor bar pass frequency is surrounded by sidebands of 2× line frequency, then the defect in the motor is:

    (a) Cracked rotor bars

    (b) Eccentric rotor

    (c) Loose stator coils

    (d) Loose rotor bars.

5.10 In a turbo machine, a sudden change in 1× amplitude and phase is an indicator of:

    (a) Unexpected fouling

    (b) Increased bearing clearances

    (c) Shaft crack

    (d) Inception of oil whirl.

6   Correcting faults that cause vibration

6.1   In a single plane balancing, if the unbalance weight is moved clockwise with certain degrees, the phase or the reference mark under the strobe moves with:

    (a) Same angle but in counter-clockwise direction

    (b) Same angle and direction

    (c) 180° from the original position

    (d) Double the angle in counter-clockwise direction.

6.2   A two plane field balancing being conducted with the conventional method will require:

    (a) 1 trial run

    (b) 2 trial runs

    (c) 3 trial runs

    (d) $N + 2$ trial runs where N is the critical speed of the rotor.

6.3   The number of correction planes depends on operating speed of the rotor. As per the rule, the number of correction planes required is (where N is number of critical speeds above the rotor operating speed):

    (a) N

    (b) $N + 1$

    (c) $N + 2$

    (d) $N + 3$.

6.4   A rotor turns at a speed of 1000 radians/s. Grinding the rotor at a radius of 350 mm achieved a balancing correction. Using the balancing standard ISO 1940, what is the balance quality if the eccentricity of residual unbalance is 0.003 mm?

    (a) G 3.5

    (b) G 3

(c) G 2.5

(d) G 0.35.

6.5 Which of the following factors usually does not affect the alignment checks of machines?

(a) Bracket sag

(b) Soft foot

(c) Axial float of shafts

(d) Electrical runout.

6.6 In the two dial method of alignment, which mathematical principle is used to calculate the shim thickness?

(a) Congruent angles

(b) Trigonometry

(c) Pythogoras' theorem

(d) Similar triangles.

6.7 A three dial alignment is done when:

(a) A complete rotation of 360° is not possible

(b) Shafts have considerable axial float

(c) Distance between shaft ends is large

(d) Diaphragm coupling is used.

6.8 In the reverse dial method of alignment:

(a) Accuracy is not affected by axial movement of shafts

(b) As both the shafts are rotated together, runouts on coupling hubs are not measured

(c) Geometric accuracy is better than two dial method

(d) All of the above.

6.9 Alignment tolerances should consider:

(a) Offset

(b) Angularity

(c) Combination of angularity and offset

(d) Vibration amplitude and offset.

6.10 While designing a dynamic absorber to resolve a resonance problem, the absorber is designed to have natural frequency:

(a) Same as that of the main mass to which it is attached

(b) Slightly less than that of the main mass to which it is attached

(c) Slightly more than that of the main mass to which it is attached

(d) None of the above.

7 Oil and particle analysis

7.1 A typical lubricant (petroleum-based) used in any machine is prepared by:

(a) Processing the crude oil

(b) Chemical reaction between low molecular components

(c) Blending the additives to a base oil

(d) Blending synthetic oil to mineral oil.

7.2   Primary oil sampling ports are used for:

(a) Detecting which component is wearing out

(b) Routine sampling

(c) Checking the condition of filters

(d) Flushing sampling bottles.

7.3   Primary oil sampling ports are typically located:

(a) On single return line just upstream of the sump or reservoir

(b) After the oil filters

(c) Downstream of the components such as bearings

(d) Downstream of oil pump.

7.4   When oil is collected for sampling from an oil sump, the best location to take the sample is from the:

(a) Oil level gage

(b) Bottom of the sump

(c) Middle level of the sump

(d) Sampling from the sump is not recommended.

7.5   As per good practice, the signal to noise ratio, which is the ratio of target oil cleanliness to maximum allowable bottle contamination, should be:

(a) 20:1

(b) 1:5

(c) 1:10

(d) 5:1.

7.6   In two consecutive reports, the ISO particle code is as follows: (1) ISO 19/12 and (2) ISO 20/13. This indicates that the particle count in the two samples has:

(a) Doubled

(b) Halved

(c) Quadrupled

(d) No relationship.

7.7   The viscosity index of an oil is the relationship between:

(a) Viscosity and acidity

(b) Present viscosity and original viscosity

(c) Viscosity and temperature

(d) Viscosity and specific gravity.

7.8   One of the following techniques is not an oil contaminant analysis technique:

(a) Timken OK value

(b) Spectrometric analysis

    (c) Ferrography
    (d) Infrared analysis.

7.9 In a particular application, the oil sample is expected to have non-ferrous particles in the size range of 5–10 microns. The most suitable analysis technique would be:

    (a) Spectroscopy
    (b) Rotrode filter spectroscopy (RFS)
    (c) Ferrographgy
    (d) Particle counting.

7.10 Identify the type of wear seen in this ferrogram (1000×):

    (a) Rolling element bearing wear
    (b) Gear scuffing wear
    (c) Black oxide due to lack of lubrication
    (d) Break-in wear.

8   Other predictive maintenance techniques

8.1 Ultrasounds are waves with frequencies over:

    (a) 20 Hz
    (b) 20 kHz
    (c) 100 kHz
    (d) 100 Hz.

8.2 One of the following is not a characteristic of an air-borne ultrasound wave:

    (a) Blocked by small objects
    (b) Does not penetrate solid objects
    (c) Radiate in a straight line
    (d) Can travel large distances.

8.3   Ultrasound can normally not be heard by the human ear, yet using a certain process it can be converted into the audible range. This process is called:

    (a) Digital signal processing
    (b) Fast Fourier transform
    (c) Heterodyning
    (d) Amplification.

8.4   The ultrasound technique is not an effective technique to detect:

    (a) Leaks
    (b) Bearing defects
    (c) Electrical defects such as arcing and corona
    (d) Fouling.

8.5   Thermography utilizes which part of the electromagnetic spectrum:

    (a) Infrared
    (b) Ultrasonic
    (c) Ultraviolet
    (d) Visible.

8.6   Thermography is generally not effective at temperatures below:

    (a) 0 °C
    (b) −20 °C
    (c) −100 °C
    (d) −180 °C.

8.7   Thermography is not very effective to detect:

    (a) Loose connections in an electrical circuit
    (b) Misalignment of a coupling
    (c) Structural resonance
    (d) Leaks in furnace brickwork.

Answers

1.1	a		2.1	d
1.2	c		2.2	d
1.3	b		2.3	0 cm
1.4	c		2.4	c
1.5	Critical		2.5	d
	Essential		2.6	c
	General purpose		2.7	a
1.6	Essential equipment		2.8	a
1.7	b		2.9	a
1.8	a		2.10	d
1.9	a			
1.10	b		4.1	b
			4.2	c
3.2	b		4.3	a
3.3	d		4.4	d
3.4	a		4.5	c
3.5	b		4.6	b
3.6	c		4.7	c
3.7	d		4.8	c
3.8	b		4.9	d
3.9	d		4.10	b
3.10	a			
			6.1	a
5.1	a		6.2	c
5.2	b		6.3	c
5.3	d		6.4	b
5.4	d		6.5	d
5.5	c		6.6	d
5.6	a		6.7	b
5.7	a		6.8	d
5.8	a		6.9	c
5.9	d		6.10	a
5.10	c			
			8.1	b
7.1	c		8.2	d
7.2	b		8.3	c
7.3	a		8.4	d
7.4	c		8.5	a
7.5	d		8.6	b
7.8	a		8.7	c
7.6	d			
7.7	c			
7.9	b			
7.10	a			

Appendix B

Practical sessions

A   Concept of natural frequency of vibration

Given

Spring; 22.8 g mass or 47.8 g mass; calculator; stopwatch; tape measure.

Assignments

1. Determine the spring stiffness by using Hooke's law, which is given by:

$$F = k \cdot x$$

   where F = force (N); x = displacement (m); k = spring stiffness (N/m).
2. Determine the theoretical natural frequency of vibration (in Hz) of the system with the given mass (ignore the spring's own weight). Natural frequency is given by:

$$\omega_n = \sqrt{\frac{k}{m}}$$

   where ω_n = frequency (rad/s); k = spring stiffness (N/m); m = mass (kg).
3. Let the spring vibrate freely with one mass and count the number of cycles in 10 s. What is the natural frequency of vibration of the system given by this experiment?
4. How does the experimental and theoretical results correlate?
5. How can we change this system to have a natural frequency of exactly 1.5 Hz?

B   The 'bump' test

Given

Aluminum beam; impact hammer; accelerometer; vibration analyzer.

Use the following settings on the analyzer

Mode: frequency; spectral lines: 800; F-max: 2000 Hz; averages: 4; average type: linear; overlap: 50%; trigger: single; source: internal; synchro-start: off.

Assignments

1. Set the analyzer to the frequency domain. Estimate the natural frequency of the beam by attaching the accelerometer in the lateral direction and conducting a 'bump' test with the hammer and vibration analyzer. Hammer in the same axis as the accelerometer.
2. Change the position of the accelerometer and repeat the test.
3. Attach the accelerometer in the sideways direction and repeat the test. Remember to hammer in the same axis.
4. Use a harder/softer tip on the hammer and repeat the test. Determine which tip is the best and state reasons why a particular tip is better.
5. Add extra damping to the beam (e.g. hold it by hand) and repeat the test. How does this influence the result?

C   Blade pass frequency

Given

Accelerometer; vibration analyzer; fan demo kit; tachometer; stroboscope (when available); calculator; presstick ('sticky putty').

Use the following settings on the analyzer

Mode: frequency; spectral lines: 800; F-max: 500 Hz; averages: 4; average type: linear; overlap: 50%; trigger: free run; source: internal.

Assignments

1. Switch the fan on (speed 1) and use the tachometer to measure the rotational speed of the fan. Use the stroboscope to verify the result.
2. Calculate the blade pass frequency (BPF) for the fan running at speed 1.
3. Use the vibration analyzer and measure the vibration spectrum in the radial direction. Identify the rotational frequency (1×) and the BFP in the spectrum.
4. Repeat the test with the axial direction measurement position. Which is the best direction for picking up the BPF?
5. Repeat steps 1–4 for speed 2 on the fan (if available).
6. Put a small amount of presstick on one of the blades and measure the vibration spectrum. Again, compare the result of the unbalanced fan with the previous results.
7. Now put presstick on two of the blades and repeat the measurement.

D   Rotor unbalance and misalignment

Given

Accelerometers; vibration analyzer; rotor demo kit; tachometer; variable power supply; presstick and shim material.

Use the following settings on the analyzer

Mode: frequency; spectral lines: 800; F-max: 1000 Hz; averages: 8; average type: linear; overlap: 50%; trigger: free run; source: tacho.

Assignments

1. Connect the accelerometers and tachometer to the analyzer. Switch the rotor on (3 V setting) and use the tachometer to measure the rotational speed. Look at the frequency domain with the analyzer.
2. Identify the rotational frequency (1×) in the spectrum and read the vibration amplitude in the spectrum at the rotational frequency.
3. Attach a small amount of presstick to the rotor and repeat the measurement. Compare the vibration amplitude at the fundamental frequency (1×) with the previous result.
4. Loosen the rotor base and use a shim to misalign the rotor. See if you can misalign the rotor to make the 2× frequency dominant (parallel misalignment).
5. For further experimentation, you can attempt the following:

   (a) Compare the result of the different measurement locations
   (b) Change the settings on the analyzer (e.g. range, windows, resolution)
   (c) Construct the averaged time signal
   (d) Explain the many harmonics of the 1× frequency.

E   Gear mesh frequency

Given

Accelerometers; vibration analyzer; gear demo kit; tachometer and calculator.

Use the following settings on the analyzer

Mode: frequency; spectral lines: 800; F-max: 1500 Hz; averages: 8; average type: linear; overlap: 50%; trigger: free run; source: internal; synchro-start: off.

Assignments

1. Connect the accelerometers and tachometer to the analyzer. Switch the rotor on (speed 2 on battery or 3 V on power supply) and use the tachometer to measure the rotational speed of the rotor. Use the schematic figure of the demo kit below to calculate the different gear mesh frequencies (GMFs).

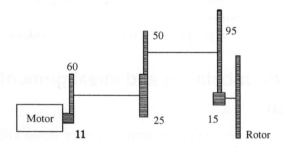

2. Use a roving accelerometer to see if you can pick up the GMFs in the system. Try to determine the best measurement locations.
3. Try to calculate the sideband spacing for each GMF and see if you can see these with the analyzer. This will require a zoom analysis (where possible).

F **Belt frequency**

Given

Accelerometers; vibration analyzer; belt demo kit; tachometer; calculator.

Use the following settings on the analyzer

Mode: frequency; spectral lines: 800; F-max: 100 Hz; averages: 4; average type: linear; overlap: 50%; trigger: free run; source: tacho.

Assignments

1. Connect the accelerometers and tachometer to the analyzer. Switch the rotor on (3 V setting) and use the tachometer to measure the rotational speed.
2. The pitch diameters and belt lengths are given on the demo kit. Use the equation on page 166 of the manual to calculate the belt frequencies.
3. Use a roving accelerometer and see if you can pick up the belt frequencies with the analyzer. Try different measurement locations. You can expect the 2× belt frequency to be dominant.
4. What may happen if a belt frequency and a rotational frequency differ with less than 10 Hz?
5. Misalign a belt with the different types of belt misalignment. See if you can pick up the change in the vibration spectrum.

Index

Printed and bound by CPI Group (UK) Ltd, Croydon, CR0 4YY

03/10/2024

01040331-0020